DESALTING SEAWATER
ACHIEVEMENTS AND PROSPECTS

OCEAN SCIENCES

Edited by **Donald A. Wilson,** *Naval Undersea Center, San Diego, California*

John E. Tyler and **Raymond C. Smith** *Measurements of Spectral Irradiance Underwater*

Marion Clawson and **Hans H. Landsberg** *Desalting Seawater – Achievements and Prospects*

Other volumes in preparation

DESALTING SEAWATER
ACHIEVEMENTS AND PROSPECTS

Marion Clawson and **Hans H. Landsberg**
Resources for the Future, Inc.
Washington, D.C.

GORDON AND BREACH SCIENCE PUBLISHERS
New York   London   Paris

Copyright © 1972 by
   Gordon and Breach, Science Publishers, Inc.
   440 Park Avenue South
   New York, N.Y. 10016

Editorial office for the United Kingdom
   Gordon and Breach, Science Publishers Ltd.
   12 Bloomsbury Way
   London W.C. 1

Editorial office for France
   Gordon & Breach
   7–9 rue Emile Dubois
   Paris 14ᵉ

Library of Congress catalog card number 76-121624. ISBN 0 677 02710 9. All rights reserved. No part of this book may be reproduced or utilized in any form or by any means, electronic or mechanical, including photocopying, recording, or by any information storage and retrieval system, without permission in writing from the publishers. Printed in east Germany.

SERIES INTRODUCTION

NATURE HAS endowed man with a legacy of untold value and potential. This legacy consists of the oceans of the world, together with its lakes, rivers, and tributaries. Man, in return, has not used this legacy wisely and is presently pursuing a course destined eventually to destroy productivity through unchecked pollution and heedless destruction of ecological systems. Each of the volumes selected for publication in the *Ocean Sciences* series will, in its selected area, help the reader to better understand this vast and complicated system and will assist the decision maker in determining national and international goals as well as regulations to better these goals.

Donald A. Wilson

Series Editor

CONTENTS

    Series Introduction v

I   FOREWORD
    Marion Clawson and Hans H. Landsberg 1

II   FIRST UNITED NATIONS DESALINATION PLANT SURVEY — A TECHNICAL AND ECONOMIC ANALYSIS OF THE PERFORMANCE OF DESALINATION PLANTS IN OPERATION
    Department of Economic and Social Affairs . . . 9
    Introduction 9
    Summary and Conclusions 10

III   1968 SALINE WATER CONVERSION REPORT
    Office of Saline Water 20

IV   SUCCESSFUL LARGE-SCALE DESALTING
    J. B. Wright 26
    Single-unit flash evaporator at Key West is producing 2.62 mgd of fresh water at a cost of 84.5¢/1000 gallons   26
    Successful large-scale desalting 30

V   A DESERT SEACOAST PROJECT AND ITS FUTURE
    Carl N. Hodges 35
    Power production 38
    Water production 38
    Food production 40

VI   1. ENGINEERING FEASIBILITY AND ECONOMIC STUDY FOR DUAL-PURPOSE ELECTRIC POWER-

	WATER DESALTING PLANT FOR ISRAEL	44
	2. EFFECT ON COSTS OF INCREASING CAPACITY TO 300 MEGAWATTS	58
VII	CONSTRUCTION OF NUCLEAR DESALTING PLANTS IN THE MIDDLE EAST HEARINGS	62
VIII	NUCLEAR ENERGY CENTERS – AGRO-INDUSTRIAL AND INDUSTRIAL COMPLEXES	
	Oak Ridge National Laboratory	93
	Prefatory note	93
	Introduction	96
	Summary	102
IX	PROSPECTS FOR DESALTED WATER COSTS	
	William E. Hoehn	122
	Introduction	122
	Costing methodologies	125
	Some case studies in problems of cost estimation: the MWD plant	132
	The Israel plant	138
	Parametric estimation of current desalting costs	146
	On large-scale desalting	154
	To sum up	168
X	THE BOLSA ISLAND DUAL-PURPOSE NUCLEAR POWER AND DESALTING PROJECT	
	Raymond W. Durante	171
	Introduction	171
	Description of the project	172
	Participants in the project	174
	Contract structure	176
	Project organization	177
	Project chronology	178
	Costs	181
	Other factors	181
	Bolsa Island project to be restudied	190

Contents

XI	NUCLEAR POWER AND WATER DESALTING PLANTS FOR SOUTHWEST UNITED STATES AND NORTHWEST MEXICO	192
	Introduction and approach	192
	Conclusions and recommendations	202
XII	PRACTICAL CONSIDERATIONS IN DESALTING AND ENERGY DEVELOPMENT AND UTILIZATION	
	James T. Ramey	207
	Introduction	207
	The case for nuclear desalting	208
	The technological base	210
	Preliminary preparations for a project	211
	Determining project viability	213
	Organizing the project	215
	Establishing a schedule for the project	216
	International cooperation	217
	Conclusion	219
XIII	REVIEW AND ANALYSIS OF THE COSTS OF DESALTED SEAWATER	
	Paul W. MacAvoy and Frederick J. Wells	221
	Summary of costs analyzed	221
	Bibliography of reference sources	235
XIV	DRY LANDS AND DESALTED WATER	
	Gale Young	238
	Water requirements for crops	240
	Production cost estimates for grain	243
	Desalination	246
	Summary	247
XV	DESALTED SEAWATER FOR AGRICULTURE: IS IT ECONOMIC?	
	Marion Clawson, Hans H. Landsberg and Lyle T. Alexander	251

Desalting seawater for large-scale commercial agriculture	254
Nuclear energy	255
Economies of scale	257
Availability of capital	257
Costs of equipment	258
Operating performance	258
Desalting process	261
Application of desalted seawater to the land	263
Value of irrigation water	265
Index	272

I

FOREWORD

Marion Clawson* and Hans H. Landsberg*

IN THE PAST TWO decades there has been a rising tide of popular and professional interest in the possibilities of desalting seawater.† Desalted seawater is now used as a source of personal, industrial, and municipal water in a number of communities around the world, especially those in arid lands located on sea coasts. The plants in use today are relatively small, all under 5 million gallons per day (mgd) capacity. (As a basis of comparison, a million gallons a day capacity would provide sufficient water for a town of perhaps 7,500 people, or would provide irrigation water for possibly 300 acres of irrigated land.) The volume of water in all the seas of the world staggers the imagination; the technologies for getting pure water out of seawater, or for extracting salt from seawater, exist; the catch has been, thus far, that the costs are too high for large-scale uses of water, particularly for irrigation. If desalting costs could be drastically reduced, the economical supply of water that could thus be opened up would solve many water supply problems in many parts of the world. Various proposals have been made to construct very large desalting plants (ranging from 100 to 1,000 million gallons per day) at costs of water which might make irrigation economically possible. It is the possibilities for the future, rather than the results to date, which make desalting of seawater so attractive.

The purpose of this book is to bring together, in one easily accessible place, a modest number of reports and articles which have appeared in a number of widely separated sources. These reprinted materials are intended to throw as much light as is reasonably possible, in one book, upon the achievements

---

* We wish to acknowledge the substantial help we have had from our colleague, Lyle T. Alexander, who collaborated with us in the article reprinted here, and whose encouragement and critical review was most helpful in the development of our knowledge of this subject.

† For a rather brief, nontechnical, and well written description of the various means of desalting seawater, see *The A-B-Seas of Desalting*, Office of Saline Water, U.S. Department of the Interior, Government Printing Office, Washington, 1968. For a book of selected reprints on the more technical aspects of seawater desalting, see Sumner N. Levine, editor, *Desalination and Ocean Technology*, Dover Publications, Inc., New York, (1968).

of desalting at the end of 1969, and to explore the hopes, problems, and prospects for the future. As economists interested in the economic use and development of natural resources, we have focused our attention on the costs and outputs of desalting processes and proposals, rather than upon their technical aspects, and have chosen reprints from the background of our interest. Technical details are important as affecting costs and output, but we have not sought to describe such technical matters for their own value.

In a field where attention is directed to future possibilities rather than to past achievements, differences of opinion or of judgment about the future are inevitable. One guiding criterion has thus been to select items for reprinting which reasonably sample the major view-points and approaches. The authors have their own judgments about the prospects for desalting, but our selection of reprints has sought to present other views, in the language of those who hold them, for as nearly complete a coverage of various positions as is possible in a single book of moderate length.

Wherever possible, we have chosen official reports or statements, and in every case have used only materials previously published. To keep the book to a modest length (and thus the price to a modest one), we have used summaries rather than the whole of several reports; the citations enable the interested reader to obtain the whole report, but reprinting of the whole would have made this book inordinately long. Several of the reprints include footnote source references for that report or article; in a few cases, bibliographies were appended in the original source, and we have reproduced those as well. In these ways, we have made it possible for the interested reader to dig more deeply into the subject now, and to keep up with developments in it over the next several years. A book dealing with a live issue of resource management unavoidably gets out of date in a few years; a book cannot serve as a newspaper, or even as a periodical journal. We hope we have presented a reasonable view of desalting seawater as of the end of 1969, but the reader will have to keep himself up to date from other sources for the future.

A brief description of the reprints included, with some of our reasons for choosing the particular reprint, may help to orient the reader. The first reprint is from a United Nations report; published in 1969, it includes data through 1965. It is thus unfortunately not as near up-to-date as one would wish, though no significant advances have in fact been made, in terms of operating facilities, in the meantime. In any event, part of its significance—aside from its being a world-wide survey—lies in the fact that this is the first of what is authorized to be an annual series of such surveys. Those interested in desalting seawater would thus do well to obtain similar annual reports for later years. The second reprint is Secretary of the Interior Stewart L. Udall's report to President Lyndon B. Johnson of the status and situation of saline water conversion programs in that Department at the end of 1968. To a considerable extent, it complements on the U.S. scene the United Nations report for the world; and, like the UN report, the Secretary's report is an annual affair which enables the interested reader to keep abreast of changing times.

The Westinghouse Electric Corporation has built a number of desalting plants. The operations of the one providing municipal water for Key West, Florida, during the past year or two are described by J. B. Wright of that company. This plant is the largest single unit actually operating in the world, according to this article; however, in comparison with some of the proposed plants, described in other reprints, which would produce from 30 to 300 times as much, it is relatively small. Helped by a loan from the federal government at 4% interest and 40 years repayment, the estimated cost of water is 84 cents per 1,000 gallons—not excessive for municipal water, though prohibitive for agriculture. In spite of some unexpected difficulties, the plant is performing about as planned, and better in several respects. It is as good an example as we could find in the literature of the small desalting plant for supplying fresh water to small communities in water-short locations.

Carl N. Hodges describes what is widely known as the Puerto Peñasco project, which produces electricity, fresh water, and vegetables for a small fishing town in Mexico on the Gulf of California. The great merit of this project is that it, like the Key West plant, is a going concern; Hodges can and does tell what has been done, rather than what it is hoped or expected might be done in the future. He expresses the opinion that similar installations might be made on other seacoast locations in the world; in fact, at least one such installation is now under construction in the Arabian peninsula near the Persian Gulf, under the technical direction of some of the same people who built and operated Puerto Peñasco. While these installations are small, not seeking to produce staple agricultural crops, yet they are most interesting. Regrettably, no firm cost calculations have yet been published either for this project or this type of installation generally.

Israel has perhaps developed its natural water supplies more fully than has any other country; future population growth will put increasing pressure on these limited supplies. Accordingly, particular interest focuses on the possibility of desalting seawater for that country. Following conversations between President Lyndon B. Johnson and Prime Minister Eshkol, a study of the engineering feasibility of a dual-purpose water-power plant was undertaken jointly by two American firms. A preliminary report was published in late 1965, a revised version in early 1966, and the latter was updated in late 1967. Our reprint is the introduction and summary from the latter two reports. The technical aspects of a plant are described, and estimates made of water costs under alternative assumptions as to interest rates and other factors. Because it dealt with Israel, a country involved in disputes with its neighbors, unusual attention has been focused on this report.

An expression of this interest in Israel and its neighbors was found in a Senate Resolution (Number 155) in 1967, and in hearings upon it. A large number of Senators sponsored the resolution, which was passed; while not binding, in terms of government action, it nevertheless sought to express the sense of the Senate. We have reprinted some excerpts from the rather lengthy hearings. Sponsors and supporters of the idea stressed its peace-making possibilities, which seemed then to careful observers as overdrawn, and in more recent months clearly can be seen to be seriously exaggerated. The proposal was advanced as economically sound, although some of its supporters hedged in their testimony on the latter point. Much was made of the Bolsa Island project in the United States, as a forerunner of what could be done elsewhere. As another reprint makes clear, this latter project is now regarded as uneconomic.

The Oak Ridge National Laboratory, with the support of the Atomic Energy Commission and the Department of the Interior, has developed in some detail the plans for an imaginative very large-scale agro-industrial complex built upon nuclear power. Its report develops an analysis in terms of near-term and far-term technologies, the latter applicable perhaps 20 years from now. The project would include electric power, desalted seawater for agriculture, and industries to use a substantial part of the power. A basic feature of the proposal is the integrated nature of these enterprises. While the proposal is general, in the sense that the complex might be located in any one of several places in the world, yet the possibility of such a complex in the eastern Mediterranean area was clearly in mind. The full report is rather long and somewhat complex, and is supported by other studies of the Laboratory; the reprint is primarily the Introduction and Summary. The report is publicly available.

William E. Hoehn of the Rand Corporation had analyzed the prospects for desalted water costs. As he clearly points out in the beginning of his report, no study today can possibly be "definitive" because of the total lack of experience with plants of the sizes and types for which most of the plans are developed. Nevertheless, he does bring a great deal of information into focus and makes many sprightly and useful observations about the probable accuracy of many of the cost estimates. His conclusions are quite unfavorable to the idea of early development of large desalting projects as a basis for agriculture. His report is clearly required reading for anyone interested in the economic aspects of water desalting, as they appeared in early 1969.

Raymond Durante describes in some detail and with care the history of the Bolsa Island proposed project in the Los Angeles area, including the reasons why it has been postponed, perhaps indefinitely. This was a moderately large-scale project, planned to produce 150 million gallons of fresh water daily, and has come nearer to being carried out than any proposal of similar magnitude to date. Its dual-purpose, electricity and water, character is described fully in the reprint. The proposal was greatly complicated by the participation of six quite diverse organizations which proposed to pool their interests in the project. The original estimated costs were $444 million, of which the federal government was to contribute $72 million; three years later, estimated costs had risen to $765 million, and some of the parties refused to proceed on the basis of these costs. Since this project is in an area where the demand for water and power is high and rising, and since a part of its cost would have been subsidized, its postponement as uneconomic is sobering, especially if one contemplates similar projects in regions of the world where such favorable demand conditions do not exist.

Some irrigated areas in southwestern California and Northwestern Mexico face a rather difficult future since Colorado River water is fully committed and may no longer be available to them in the future in the amounts they have enjoyed in the past. Hence, interest in desalting seawater is high in both countries. A preliminary assessment of the possibilities and problems has been made; its introduction and conclusions are reprinted here. This area is a good example of a situation where desalting seawater would be a valuable solution to a difficult problem, if costs can be brought down to levels that the local economy can tolerate.

James T. Ramey, a Commissioner of the Atomic Energy Commission, has shown much interest in the potentialities of desalting seawater, and represents an agency with a lively interest in the subject. His paper, given near the end of 1968, is a strong statement of the affirmative size of this type of development. At the same time, he discusses some of the problems that will be encountered and that must be solved, if such projects are to go forward.

As comprehensive a review of costs of desalting seawater as the data permitted has been prepared by Paul W. MacAvoy and Frederick J. Wells. As they point out, cost estimates differ widely, in part due to differences in ways of calculating them. Their report attempts to put a considerable number of cost

1 Clawson/Landsberg (0271)

estimates on a reasonably comparable basis, and to evaluate the quality of some estimates. As they point out, there is remarkably little actual experience to go on, even for small plants, and none whatsoever for the large plants; one must use estimates of future costs, and these may well turn out to be too optimistic—as they did for Bolsa Island. The authors make their own estimates of desalting costs, under a variety of conditions, but based upon the estimates made by others. To anyone interested in the cost of desalted water, this report is surely a basic document, both for its comprehensiveness and listing of a number of studies (which list has been included in the reprint), and its analytical detail.

Gale Young, Assistant Director of the Oak Ridge National Laboratory, has written an explanation of the Oak Ridge and other proposals for large scale desalting of seawater for staple crop agriculture, in which he takes a far more optimistic view than do some others. He deals with management of the land and water, and the costs of production, of the crops that could be grown. He also dwells upon the attractiveness of coastal desert regions as places for living; here, desalted seawater might play a particularly useful role.

Finally, we include our own article, with our analysis of the prospects for desalting seawater for commercial agriculture. As economists, our approach is more cautions than that of the advocates of large-scale desalting projects; we think there is no prospect that such plants can be economically sound for commercial agriculture during the next 20 years.

We present these reprints without further comment, urging readers to read each carefully, and to compare one with another. A man or a group may, in all honesty and conviction, estimate a cost or a return from a proposed plant which in the judgment of others is unrealistic. Merely because some report states that costs are estimated at some figure does not prove that they will in fact be at that level or indeed that this is a reasonable advance estimate. In a field where so much is speculative and so little is proven by experience, there is room for many differences of opinion among equally competent and well-intentioned workers. This is especially true for the largest planned plants, which have yet to be tested by experience. Likewise, the reviewer's criticism may prove in error and should be carefully weighed by the reader.

Although the different reprints vary greatly among themselves in many respects, and hence are not easily classified, perhaps a rough grouping of them will prove useful to the reader. With the caution, that not everyone in each group closely resembles others in the same general group, the various reprints may be divided into three major classes, as follows:

1) The describers or explainers, who seek primarily to provide information about some experience or situation. These groups are generally favorable to seawater desalting, but they are not primarily advocates. We would put the excerpt from the UN report in this category, for it seeks to provide information on the status of seawater desalting around the world without trying to promote it. The Department of the Interior report is also grouped here, though it is somewhat more eager to promote desalting. The reports by Wright and Hodges, on actually operating plants, can also be put in this category; though each writer is sold on the success of the plant he describes, their primary thrust is not promotion. Likewise, Durante describes the Bolsa Island experience; he is clearly unhappy that it turned out as it did, but he tells fully and matter-of-factly what happened. In thus assigning a somewhat neutral role to this group, we do not exclude it from the kind of critical review which we urged upon our readers earlier; merely because a report quotes a figure or expresses a judgment does not prove that it is correct. In particular, cost figures must be regarded with care, partly because uniform treatment of costs is lacking.

2) The proposers or the advocates or the promoters, men or groups who advance ideas about possible future seawater desalting plants of a size, type, and method of operation which has no counterpart in any plant today, but which they think has great potential for the future. We would place in this group the Kaiser report on a possible plant for Israel, the Oak Ridge Laboratory proposal for an agro-industrial complex, the proposal for the U.S.-Mexico plant on the Gulf of California, Commissioner Ramey's speech, most of the discussion at the Senate Hearings, and Gale Young's article which is largely based on the Oak Ridge proposals. Some of these proposals, such as that of the Oak Ridge Laboratory, intentionally are based upon technology not yet available; they are trying to visualize what might happen at some future date. We endorse the idea of such bold and imaginative forward planning and speculation; the difficulty comes when some parts of the public accept it as demonstrated and tested fact. The fullsome praise of the Bolsa Island project in the Senate Hearings, followed by an announcement about six months later that the project had to be abandoned because the costs of its planned output were too high, should be sobering to anyone who advances bold but untested proposals.

3) The critics, persons who see the great possibilities of desalted seawater as a means of meeting serious water problems in many parts of the world, but are highly dubious about the economics of the various proposals. The

critics generally accept the statements of the foregoing groups about technical processes, but question costs, output, performance and value of the water that may be produced. We include in this group Hoehn, MacAvoy-Wells, and ourselves. Their positions are set forth in the respective reprints, and need not be repeated here. For ourselves, we are particularly concerned that developing countries not be led to invest scarce capital in projects whose economic output is as dubious as we think large-scale desalting of seawater is in 1970 and will be for the seventies and into the eighties. We are apprehensive of unexpected costs and difficulties, and doubt that there may be unexpected profits and gains to offset them. At the same time, at least the present authors, and perhaps the others in this group also, would endorse the building of one moderately large plant to test these proposals in operation—but under a clear understanding by all concerned that it was an idea as yet unproven, rather than an economically viable enterprise.

With these comments, we leave the reprints to speak for themselves. We believe they provide expression of all major viewpoints and enable the interested reader to go further if he wishes. If we have served to inform and to stimulate, but perhaps not to satisfy, then our objective has been fullfilled.

II

FIRST UNITED NATIONS DESALINATION PLANT SURVEY–A TECHNICAL AND ECONOMIC ANALYSIS OF THE PERFORMANCE OF DESALINATION PLANTS IN OPERATION[1]

Department of Economic and Social Affairs

United Nations, New York

Introduction

THE FIRST *United Nations Desalination Plant Operation Survey* is part of the research and studies programme in the field of desalination undertaken by the Resources and Transport Division of the Department of Economic and Social Affairs of the United Nations Secretariat. These studies pay particular attention to the application of desalination technology in developing countries. Among the previous publications in this field reference should be made to *Water Desalination in Developing Countries,* Water Desalination: *Proposals for a Costing Procedure and Related Technical and Economic Considerations, Proceedings of the Interregional Seminar on the Economic Application of Water Desalination* and *The Design of Water Supply Systems Based on Desalination.**

In the course of the United Nations Interregional Seminar on the Economic Application of Water Desalination, which was held at United Nations Headquarters from 22 September to 2 October 1965, participants expressed appreciation of the programme of round-table discussions at which the actual performance of a number of desalination plants already in operation could be reviewed. This provided an opportunity for discussion of the difficulties encountered in plant operation and the remedies applied in each case. Among the recommendations formulated by the participants in the Seminar was the suggestion that the United Nations Secretariat should prepare a basic ques-

---

* United Nations publications, Sales Nos.: 64.II.B.5, 65.II.B.5, 66.II.B.30 and E.68.II.B.20, respectively.

tionnaire to be sent at regular intervals to authorities charged with the responsibility of operating desalination plants, in order to assemble and maintain up-to-date operational data on desalination plants, which should be published and distributed periodically. Such a publication should be a useful source of information in the field of desalination, particularly for those developing countries that are interested in the application of water desalination.

These recommendations were included in the report of the Secretary-General, *Water desalination with special reference to developments in 1965* (E/4142)† and were approved by the Economic and Social Council in resolution 1114 (XL) of 7 March 1966.

The Secretary-General wishes to acknowledge on this occasion the valuable co-operation rendered by the Government of the United Kingdom which, following the recommendations of the Economic and Social Council in resolution 1114 (XL), has provided under a funds-in-trust programme the services of desalination experts to assist the Secretariat in the implementation of its work programme in the field of water desalination.

Summary and conclusions

Scope of survey

The present survey is the first of a continuing series planned with the objective of compiling a progressive record and analysis of the economic and technological development of desalination plants throughout the world.

The survey includes all plants engaged in commercial production for which it was possible to obtain records of operation throughout 1965. Plants with a capacity of less than 10,000 gallons per day specialized boiler feed-water plants and experimental plants were excluded.

The survey has been made possible through the co-operation of the many authorities responsible for the operation of desalination plants. Data have been obtained from questionnaires completed by these authorities. It should be emphasized, however, that the operating authorities are not necessarily responsible for the interpretation given to these data within this report.

Eighty-seven plants in twenty-one countries were surveyed. Data were

---

† *Official Records of the Economic and Social Council, Fortieth Session, Annexes*, agenda item 7.

obtained on the multi-flash, long-tube vertical, submerged-tube and vapour-compression distillation processes and the electrodialysis process. The total productive capacity of these plants is 24,994,000 gallons per day. Table 1 shows the geographical distribution of the plants; table 2, the capacity distribution; table 3, the year of construction; and table 4, the application.

Multi-flash distillation plants account for more than two-thirds of the total production capacity of plants of all sizes built since 1957 throughout the world.

Although sub-merged-tube distillation plants are as numerous as multi-flash units, they account for less than one fifth of the latter group's capacity. Nearly all of the submerged-tube distillation plants were installed in Kuwait between 1950 and 1955, and were built in groups of eight or ten because of technological limitations on unit capacity. By current standards, the submerged-tube distillation process is properly considered obsolete. Indeed, in 1968—about fifteen years after their construction—the Kuwaiti authorities initiated a programme of scrapping these submerged-tube units.

The inherent limitations of the submerged-tube distillation process have been circumvented in the related long-tube vertical distillation process now being developed to compete with multi-flash distillation. One long-tube vertical distillation plant is included in this survey.

While the vapour-compression distillation process is widely used, it is mainly for small-capacity units.

Electrodialysis, which is a membrane process using electricity for operation, is the only non-distillation process in the survey. At the current time, economic considerations limit the practical application of electrodialysis to brackish-water conversion. The process is more widely used for small plants than is indicated by the data. Although electrodialysis is now rapidly advancing to the stage of widespread commercial application, at the time of the survey the process was still undergoing intensive development towards this larger scale application.

The complementary requirements of high-quality heat for power generation and low-quality heat for distillation lead to the consideration of dual-purpose power/water plants, which offer the possibility of a reduction in over-all cost. Thus, two thirds of the multi-flash distillation plants are integrated with power production, as were all the preceding submerged-tube distillation units.

A unique example of dual-purpose operation is the combination of water production with salt production.

The range and distribution of process types included in the survey provide

Table 1 Geographical distribution of plants surveyed, by process.

	Number of plants				
	Distillation				
Location	Multi-flash	Long-tube vertical	Submerged tube	Vapour compression	Electro-dialysis
North America					
United States of America	1	1	–	4	5
South America					
Ecuador	1	–	–	–	–
Peru	–	–	–	1	–
Venezuela	1	–	–	–	–
Caribbean					
Bahamas	1	–	–	2	–
Bermuda	1	–	–	1	–
Cuba	3	–	–	–	–
Leeward Islands	–	–	–	1	–
Netherlands Antilles	3	–	–	–	–
Virgin Islands	2	–	–	–	–
Europe					
Finland	–	–	–	–	1
Gibraltar	2	–	–	–	–
Italy	1	–	–	–	–
Africa					
Libya	3	–	–	–	1
Asia					
Abu Dhabi	1	–	–	–	–
Das Islands	3	–	–	–	–
Kuwait	9	–	28	–	–
Japan	–	–	–	1	–
Pacific					
Marshall Islands	–	–	–	6	–
French Polynesia	–	–	2	–	–
Atlantic					
Ascension Islands	–	–	–	1	–
TOTAL	32	1	30	17	7

Table 2 Plant capacity, by process (Thousands of U.S. gallons per day).

| | Number of plants ||||||
| | Distillation ||||
Capacity range	Multi-flash	Long-tube vertical	Submerged tube	Vapour compression	Electro-dialysis
10–29	–	–	2	11	2
30–99	12	–	–	2	3
100–299	4	–	28	2	1
300–999	9	–	–	1	1
1000–1999	7	1	–	1	–
Total number of plants	32	1	30	17	7
Total capacity	17,196	1,000	3,392	2,446	960
Grand total capacity			24,994		

Table 3 Plant construction, by process, 1950–1965.

| | Number of plants |||||
| | Distillation ||||
Year	Multi-flash	Long-tube vertical	Submerged tube	Vapour compression	Electro-dialysis
1950	–	–	8	–	–
1951	–	–	–	6	–
1952	–	–	–	–	–
1953	–	–	10	–	–
1954	–	–	–	–	–
1955	–	–	10	4	–
1956	–	–	–	1	–
1957	4	–	–	1	–
1958	1	–	–	–	1
1959	1	–	–	–	1
1960	4	–	–	–	1
1961	2	1			1
1962	8	–	–	–	1
1963	4	–	2	1	–
1964	6	–	–	3	2
1965	2	–	–	1	–
	32	1	30	17	7

Table 4 Plant application, by process.

	Number of plants				
	Distillation				
Application	Multi-flash	Long-tube vertical	Submerged tube	Vapour compression	Electro-dialysis
Water only	9	1	2	16	7
Water and power	20	–	28	–	–
Water and salt	–	–	–	1	–
	29	1	30	17	7

a useful representation of the practical status of desalination as it existed up to 1965. Predictably, no commercial plants were reported using the reverse osmosis (hyperfiltration) or freezing processes. Future surveys are likely to show a shift in this balance. Indeed by 1968, there had already been a steady advance in the accepted upper limit to the size of multi-flash distillation units, while the near future will undoubtedly see the development of, initially small, commercial reverse-osmosis units and possibly some freezing units. Apart from the intrinsic merits of competing processes, however, local circumstances will continue to play an important part in the selection of particular plant configurations.

Major findings

The survey confirms desalination as an entirely viable source of water-supply, at least from the point of view of technical feasibility, under conditions where proper attention is given to the design, construction and operation of both the plant and the water-supply system of which it forms a part. More specifically, it is evident that the design and construction of the plant must be soundly based: adequate operation and maintenance personnel and facilities must be available; and the over-all water-supply system must be so designed that demand can be sustained at all times despite the occasional stoppages to which the desalination plant may be subject and despite seasonal and long-term variations in the pattern of demand. These last considerations result in a reduction in plant load factors to cover maintenance periods and the need for storage or spare desalination capacity to sustain supply during these periods. Similarly, load factors are further reduced by the need to install desalination capacity sufficient to meet seasonal peaks and long-term growth in demand.

In respect of desalination costs, the survey shows the extreme variation between the water costs for different or similar desalination plants operated under various conditions; emphasis is thereby given to the hazards of attempting any generalized statement on these costs. It is evident, however, that the water costs as reported here for actual installations are distributed at a considerably higher level than the cost estimates attached to most published design studies.

The important components of total water costs are: capital and other associated fixed charges; fuel and electricity; labour; maintenance; and chemicals. The factors influencing the capital charges cost component are: the specific investment committed in the plant; the annual interest payable on this investment and the period of amortization of the debt, together with any related capital charges such as insurance or taxes (in total, constituting the annual fixed charges); and the annual production of water over which these fixed charges can be distributed (determined by the plant load factor). It is the variability of these factors which results in the wide range of reported water costs and which, in any particular case, must be specified before meaningful determination of water cost can be made.

The more detailed findings of the survey are summarized below.

Capital investment

Capital investments for the plants surveyed are shown in Chapter 1, Table 5. The wide range of specific costs is partly due to the variable degree to which ancillaries and civil work may be included in any contract and the peculiar commercial pressures to which construction bids are subject during the years which precede the stabilization of a world market in any emergent technology.

Despite these factors, the existence of a strong trend towards reducing unit costs in larger plants is firmly established; these economies of scale are well illustrated in table 6.

Capital charges

Capital charges also are discussed in Chapter 1. Tables 8, 9 and 11 show the range of interest rates and amortization periods (expected lifetimes) applied to plant investments by the various operating authorities. The sample is strongly biased towards low rates—even zero in one notable example in the Middle East. Plant lifetimes are most frequently assessed at fifteen to twenty years, although these are necessarily forward estimates.

The resulting fixed charges are predominantly within the range of 6–10 per cent per annum.

By contrast, the report (see table 10) recommends that 10 per cent should be considered the minimum rate of fixed charges to be applied under normal economic conditions.

Load factors

The average annual load factor, indicative of the degree to which actual annual water production approaches the level achievable under continuous operation of the plant at its rated capacity, is found to vary significantly between process types. The modern large multi-flash plants have an average annual load factor of 61 per cent; the older submerged-tube units operate at an average annual load factor of 55 per cent; and the vapour-compression distillation and electrodialysis plants have average annual load factors ranging as low as 40 per cent.

Almost half of all the total number of plants operated with annual load factors between 40 per cent and 60 per cent annual load factors and two thirds, between 30 per cent and 70 per cent. (See tables 12, 13 and 14.)

The generally low load factor in the actual operation of most plants illustrates an important departure from the design ideal. While this is partly explained by plant down-time arising from maintenance requirements, of equal importance is the mismatching of plant capacity and water demand which is inevitable under conditions of seasonal and long-term demand variations. The influence of these two types of demand variations is determined by referring to the peak-month load factor and to the ratio of annual/peak load factor. The peak-month load factor is indicative of the degree to which the plant capacity is fully utilized and hence of the degree to which provision has been made for future long-term growth in demand; the ratio of annual/peak load factor is a simple measure of the seasonal variability of actual demands. (See tables 12, 13 and 14.)

Cost analysis

Water costs   Tables 15 and 16 show total and component water costs for all plants reported. Total costs, which are collated in table 17, show the extreme variability experienced. Significant cost ranges are between $1 and $2 per 1,000 gallons for one third of the plants and between $3 and $4 per 1,000 gallons for another one third of plants. Only 5 per cent of the plants achieved production costs below $1 per 1,000 gallons, while at the other extreme, 5 per cent of the plants had production costs exceeding $20 per 1000 gallons.

Component costs The distribution of component costs—capital charges, energy, labour, maintenance and chemicals—is shown in table 19, while the average distributions for the different process types are shown in table 20. The most dominant element is capital charges, which, on average, acount for about 35 per cent of the total water cost. This can be attributed to the unusually low load factors mentioned above and suggests an incomplete realization of this factor at the planning stage, resulting in some over-capitalization in plant design.

Energy costs In the multi-flash distillation plants, energy is the next most significant cost, averaging 26 per cent of the total. In the submerged-tube units, energy represents the unusually low fraction of 10 per cent of the total cost, although this is largely due to somewhat unrepresentative local conditions. Proportionate energy costs are surprisingly low for the vapour-compression and electrodialysis plants.

Labour costs Labour costs are found to be disturbingly high; on average, they account for about 25 per cent of the total water cost. The subject is discussed in some detail in the report (see chapter 3, section E, and table 30), where it is shown that the continuous manning of a plant with a capacity of 1 million gallons per day at even the low level of one operator per shift can be a significant economic burden. These findings point up the great incentives for more automatic operation of desalination plants, or perhaps more realistically, the advantages of locating desalination units with, say, a power plant so that manpower and other facilities may be shared by the two functions.

Maintenance costs By far the highest maintenance costs, often exceeding $1 per 1,000 gallons, are found in the smaller plants. Probably of greater interest are maintenance costs in the larger multi-flash units, where costs of the order of $0.10 per 1,000 gallons are experienced. Even at this lower figure, maintenance costs prove to be somewhat higher than the planning estimates provided in many studies.

Dual-purpose plants

Of the thermal distillation plants surveyed, forty-eight are dual-purpose units incorporated with power plants; only twelve are single-purpose plants, including several which are non-typical (see table 21). While the usual practice of taking low-pressure exhaust steam to drive the distillation plant in a dual-purpose installation undoubtedly offers economies, it does demand careful planning of the system to avoid the serious problems arising from a short- or long-term mismatching of power and water demands.

The location of a distillation plant and a power plant on a common site does, however, offer considerable advantages even where such close coupling of the steam-supplies is not practised. These advantages are that the two functions can share the same management, operation and maintenance facilities; in general, the requirements of the two functions are similar in these three respects. For distillation plants of anything but unusually large capcity, it is only on this "shared" basis that the rather specialized management and maintenance requirement and the part-time operating labour demands, can be met at acceptable cost.

Distillation condenser design

Distillation condenser tube specification are listed in table 26, the large element of total capital cost attributable to condensers making them of special interest. The recovery section, accounting for the most significant part of the total heat-exchange surface in the high-performance multi-flash units, are most commonly tubed in aluminium brass. There is, however, a growing tendency towards the use of one of the more expensive cupro-nickels in the heat input and rejection stages where more corrosive conditions are encountered. For the same reasons, a cupro-nickel tube material may be employed in the initial recovery stages where the brine first flashes with the release of residual dissolved gases. The great majority of multi-flash plants employ tube diameters of between three-quarters of an inch and one inch; tube thicknesses are almost universally 0.046 to 0.049 inch (18 gauge).

Scale control

For scale control, the majority of plants employ phosphate dosing, which is effective up to temperatures of about 200 °F (see table 27). This apparent preference for phosphate dosing, as opposed to acid treatment, which allows operation up to approximately 250 °F, is most probably due to two factors: most of the plants were constructed before acid scale-control techniques had been full developed; and many of the plants are located in areas where acid prices are prohibitively high.

The performance of the phosphate compounds under differing conditions of sea-water feed remains somewhat variable and even unpredictable; excellent performance has been obtained in some areas, while in others there has been a continuing history of problems, e.g., excessive sludge formation. Operating experience of some of the proprietary scale inhibitors recently developed should be awaited with interest.

Water quality and temperature

Product-water qualities and temperatures are given in table 28. A marked difference exists between distillation, which usually yields a product of 50 ppm or better, and electrodialysis, where purities are usually set at about 500 ppm – higher purities being achievable only at increased cost. With the high product purities produced in distillation units, chemical dosing or blending is usually required to ensure that the product is non-corrosive to the distribution system and to improve palatability. Where brackish water is available, this blending can result in a useful increase in water production.

The distillation process does, however, suffer the disadvantage of producing a rather high-temperature product. Actual temperatures are influenced by local sea-water temperatures and the specific plant design. In most cases, distillation product-waters were reported at around 100°F, or higher.

Operational problems

The principal causes of plant shut-down are listed in chapter 3, section F of the report. More than half of the plants reported shut-downs due to the pumps and drives, and, similarly, more than half of the plants reported troubles arising from corrosion. A significant number of plants reported blockages or fouling due to inadequate screening of the sea-water feed.

Over two-thirds of the plants reported scale problems as the cause of shut-downs.

Storage capacity

The storage capacities associated with desalination plants are listed in table 31. Some ambiguity necessarily arises in assigning these storage facilities when a number of plants are located at a common site. Taking into account these multiple-plant complexes, the majority of desalination installations are reported to have storage capacities of between one week and one month of production capacity.

Reference

1. Department of Economic and Social Affairs, United Nations, New York. ST/ECA/112, Sales No.: E.69.II.B.17. Reprinted portion is Introduction, and Summary and Conclusions.

III

1968 SALINE WATER CONVERSION REPORT[1]

Office of Saline Water

Department of the Interior, U.S. Government Printing Office

DEPARTMENT OF THE INTERIOR,
OFFICE OF THE SECRETARY,
Washington, D.C. 20240, January 17, 1969

DEAR MR. PRESIDENT: In compliance with Public Law 448, 82d Congress, 2d session, as amended, I am submitting this summary of 1968 ativities in the desalting of sea and brackish waters, together with recommendations for future legislation. More detailed information will be contained in the 1968 Saline Water Conversion Report now in the process of being published.

I am proud to report that the desalting program has produced steady progress in the development of economical methods for the production of fresh water from saline sources. It has contributed measurably to this country's desalting technology and has encouraged the growth of an industry destined to play an increasingly important role in the quest for potable water.

The growing practicability of desalting is underscored by the fact that 627 desalting plants are now providing some 225 million gallons of fresh water daily for cities and industries around the world. The total production should reach 1 billion gallons-per-day by 1975 as the demand for fresh water increases and the desalting potential is more widely recognized and accepted by water planners.

These statistics, which reflect the technological advancements we will review in this letter, can be traced directly to the cooperative effort between Office of Saline Water programs and private industry's research and engineering, as well as the growing attention being given to desalting by the governments and industries of other countries.

There can be no cooperation without communication. Accordingly, in a new approach to promoting a closer relationship and better understanding between the desalting industry and OSW, representatives of each held a series of meetings during 1968. They were rewarding in every respect. We consider such a free exchange of ideas and information an essential part of the overall effort toward achieving our desalting goal of low-cost fresh water.

When significant new information and data are obtained from research studies sponsored by OSW, the results are published and disseminated throughout the world. About 400 such reports have been issued and they are currently being released at the rate of about two reports per week.

As you know, in 1966 we obtained authorization to construct the world's first dual-purpose nuclear power and water desalting plant on a manmade island off the coast of Southern California in cooperation with the Metropolitan Water District, Southern California Edison, San Diego Gas & Electric Company, and the City of Los Angeles Department of Water and Power. It is very questionable that this project can proceed at the present time. The utilities have withdrawn their support and MWD has decided to proceed on the project at a later time.

Under the new timetable, construction would start in the mid-1970's and the plant would begin producing 50 million gallons of desalted water daily about 1980. The previous timetable had called for the start of construction in 1968, initial operation in 1974 and expansion of the plant to a capacity of 150 mgd by 1978. We are investigating other opportunities for reaching this historical milestone in the application of desalting technology.

Under an agreement between the U.S., Mexico and the International Atomic Energy Agency signed in 1965, the use of very large dual-purpose desalting plants in the Southwest was given detailed consideration. The study team considered for the first time the total water needs of a vast arid region and the potential of desalting to provide fresh water on such a scale. The study firmly established the technical feasibility of nuclear power and water desalting plants for the arid regions of California and Arizona in the U.S. and Baja California and Sonora in Mexico.

In order to take advantage of the economics of large-scale equipment, the team selected water-power plants producing 1 billion gpd of fresh water and 2000 megawatts of electricity as the basic unit size. The first plant could be on-stream in the 1980's at a site such as the El Golfo de Santa Clara area, near Riito, or on the U.S.-Mexico border near San Luis Rio Colorado. The construction of a series of these plants would produce a new river of fresh water to satisfy the needs of one of the fastest growing regions of the United States.

As the technology advances toward such large-scale desalting, costs by 1972 could be reduced to 50 cents per 1,000 gallons for the 1 to 10 million gpd plants, according to OSW projections. Within the next five to ten years, sea water desalting costs could drop to about 25 to 35 cents per 1,000 gallons with plants of 50 to 150 million gpd capacity. Sometime beyond 1980 the costs may become sufficiently low to be competitive for use in the irrigation of high-valued crops.

For comparative purposes, you are reminded that sea water conversion is now being accomplished for 85 cents per 1,000 gallons at Key West, Florida, the first city in the U.S. to turn to the salty sea for its regular municipal supply of fresh water.

I take great pride in reporting that the workhorse of the desalting business, the multistage flash distillation process, has reached new heights of efficiency since the 1 mgd Clair Engle plant was placed into operation at San Diego. We expect even greater efficiency with completion of additional facilities now under construction there. A new high-temperature unit, coupled with a pretreatment system will enable engineers to operate the Clair Engle plant at temperatures of up to 350 degrees Fahrenheit. This is about 100 degrees higher than heretofore achieved in a multistage flash operation. The modified plant, operating at this higher temperature, is expected to increase the fresh water output of the plant by 25 per cent.

The newest desalting unit at the San Diego Test Facility—a multistage flash distillation module—is the world's first experimental plant designed to provide actual construction and operating data for much larger plants.

The module consists of only a portion of a complete 50 million gpd plant, but all of the equipment and components are full size. Thus, a 78,000 gallons per minute brine recirculation pump provides full hydraulic characteristics of a complete plant and each flash stage is full size. It has been designed to provide maximum experimental flexibility. Further, in order to permit investigation of different operating conditions, field modifications and adjustments can be made to the module. For example, it can be operated as the high or the low temperature end of a complete plant.

Depending upon whether the high or low temperature method of operation is selected for test, the actual fresh water output of the plant ranges from 2.6 million gallons per day to 3.2 million gpd. The amount of water produced by the plant is incidental—the real product is engineering data. With this data, we hope to bridge the technological gap between present plants in the 2.5 million gpd range and the projected units of 50 million gpd or more of the next decade.

The reverse osmosis process, which utilizes membranes to demineralize brackish waters, is commanding ever greater attention and support. I think

1968 Saline Water Conversion Report 23

it is apparent from the remarkable advances that continue to accrue that reverse osmosis will play a dominant role in providing an incremental source of fresh water.

At this point, the rapid development of RO has enabled OSW to begin field test operations of several small mobile pilot plants designated to obtain performance data on various types of brackish waters. This data will provide guidelines for future activities in the continued development and practical application of these desalting systems. Continued progress may extend the process as a method of desalting sea water. Meanwhile, RO is being tested for possible use in depolluting irrigation return flows, recovering potable water from municipal sewage and in raw water treatment. As an integral part of this membrane development, OSW has awarded a $1,174,400 contract for construction of a Brackish Water Test Center at Roswell, New Mexico. This new facility will enable OSW to expedite experimental work on processes which show economic promise of improving the water supplies of many American cities and communities now using substandard water.

Reverse osmosis is now in full-scale commercial development. According to Stanford Research Institute, separation processes using membrane techniques will climb in the U.S. marketplace from a current $15 million to about $75 million by 1975.

In the area of materials, a development and testing program continues to search for cheaper alloys that would provide high heat transfer coefficients and, at the same time, resist the corrosive attack of hot brine. Hot sea water is a very corrosive liquid which is a continuing problem in all distillation equipment. The cost of materials alone may account for about 40 per cent of the total expenditure for desalted water.

A new Materials Test Center is under construction at the Freeport, Texas, Test Bed Facility. The center will supply sea water for six test units in subjecting such materials as copper, aluminum and nickel to corrosion. It is hoped these experiments will find the best materials for building economical plants and thus lowering the cost of desalted water.

The use of concrete is being studied and evaluated as a construction material for evaporator shells with considerable promise. Higher and higher heat transfer coefficients are being obtained from fluted, double-fluted and spiral tubes. Plant geometry is being optimized. Operating procedures are being improved. Scale control methods are being improved and operating temperatures are reaching higher and more efficient levels.

These developments—and others too numerous to list in this letter—are not "break-throughs" in the classic sense, but in the aggregate they offer substantial advances in distillation technology and point the way to lower-cost product water.

Although no one can predict the eventual magnitude of the industry, desalting already has provided water planners with an alternate solution to water supply problems. What has been established, in fact, is to establish the maximum price that man must pay for water along the world's sea coasts.

Commenting on desalting's future role, the U.S. Water Resources Council said in a 1968 study:

> As water requirements expand and supplies become scarce and more costly, alternatives such as desalting will become more attractive. Prospects for increasing efficiency of desalting technology and declining costs hold enough promise that this source of water supply should be given careful consideration in comprehensive planning for areas with a deficiency of fresh water. Research and development should continue to be given strong Federal support.

In regard to this support, we have submitted to the Congress a bill "To authorize appropriations for the Saline Water Conversion Program for Fiscal Year 1970, and for other purposes".

The Act of July 3, 1952, as amended (42 U.S.C. 1951 et seq.), authorizes "to be appropriated such sums, to remain available until expended, as may be specified in annual appropriation authorization acts". In order to meet Fiscal Year 1970 program requirements, we propose that our $27 million authorization bill be enacted so that appropriations can be considered and passed by Congress.

The authorization for FY 1970 will enable us to conduct a research and development program to meet the goals of four major activities:

1. Research and development operating expenses, $18,095,000.
2. Design, construction, acquisition, modification, operation and maintenance of saline water conversion test beds and test facilities, $5,355,000.
3. Design, construction, acquisition, modification, operation and maintenance of saline water conversion modules, $1,450.000.
4. Administrative and coordination, $2,100,000.

We also propose to amend Section 8 of the Saline Water Conversion Act (42 U.S.C. 1958) to allow limited participation in international research and development on desalination. Section 8, as amended by Public Law 90–297, now prohibits after July 1, 1968, the making of any new commitments for cooperation with public or private agencies in foreign countries which require the expenditure of funds under the Act.

Much valuable information has been gained through research and development contracts entered into with foreign research institutes, universities and individuals prior to July 1, 1968. This source of knowledge and technology should not be shut off.

Our proposal would specifically prohibit our contribution of funds for, participation as an agent for, or supervision of, the construction or operation of a foreign desalination plant or its components, but would not prohibit the use of funds for other foreign activities authorized by the Act. It would clarify our authority to engage in some foreign activities on a limited scale, while preserving the specific ban which we understand the Congress intended.

In conclusion, I should like to state in this my final report on the saline water conversion program, that while I am disappointed that we were unable to proceed with the program for a huge dual-purpose nuclear power and water desalting plant in Southern California, I am pleased to report that during my stewardship of the Department of the Interior the Office of Saline Water has achieved substantial and rewarding progress in the quest for low-cost desalted water.

I congratulate the scientists and engineers, both those in the Government as well as those in universities and industrial firms who have worked diligently to this end. We are deeply grateful to you, Mr. President, for the leadership and personal attention you have given the desalting program Progress made in desalting during your administration, Mr. President, will ensure the water supply for people tomorrow and the day after tomorrow.

We also are mindful and appreciative of the continuing support of the Congress which has enabled us to expand and accelerate the program to hasten the day when the benefits of low-cost water from the sea will be available to all mankind.

Sincerely yours,

STEWART L. UDALL
Secretary of the Interior

THE PRESIDENT,
THE WHITE HOUSE,
Washington, D.C. 20500

Reference

1. Office of Saline Water, Department of the Interior U.S. Government Printing Office, Washington. Reprinted portion is letter from Secretary of the Interior Stewart L. Udall. Dated January 17, 1969 to President Lyndon B. Johnson.

IV

SUCCESSFUL LARGE-SCALE DESALTING[1]

J. B. Wright

Westinghouse Electric Corporation

Single-unit flash evaporator at Key West is producing 2.62 mgd of fresh water at a cost of 84.5¢/1000 gallons

IN FEBRUARY 1969, the new desalting plant at Key West produced its billionth gallon of fresh water. Compared with the design specification of 2.62 mgd output, the plant reached 2.72 mgd during performance tests, making it the largest single-unit, single-purpose flash evaporator operating in the world today.

Over 205 gal of fresh water are produced for every gallon of bunker C fuel oil burned, reflecting the very low heat requirement of 68 Btu per lb of distillate. Product water-to-steam ratio is 14 lb of product per lb of steam used.

The original feasibility studies of the plant predicted a water cost of 92¢/1000 gal for full capacity operation and a 95% on-stream factor. The final proposal for operation and management of the facility predicted a cost of 84¢/1000 gal of product. Operating records over the past several months confirm the actual cost of water at 84.5¢/1000 gal. Total plant cost was slightly more than $4 million.

Performance of this plant and the fact that 47,000 residents of Key West are dependent on this source of fresh water make a strong recommendation for the proposed dual-purpose plants with higher output and lower cost.

Background and development

Design, startup and operation of the Key West plant will be discussed, but these will have more meaning if the events and studies which led up to the plant are described. In brief, Key West, Fla. is a community of about 47,000 people in the southernmost region of the Keys. It is surrounded by salt water and traditionally desperate for a reliable source of fresh water.

The first positive source of water for the Keys came in 1941 when the Navy constructed an aqueduct from Florida City to Key West. The installation consisted of an 18-in.-diam steel pipeline, a softening plant and two booster stations. By agreement with the Navy, the Florida Keys Aqueduct Commission could purchase excess water over the Navy's requirements.

By 1962, civilian requirements began to exceed this available excess, and the Navy installed three additional booster stations. These were paid for by the Commission, and the added operation and maintenance costs were covered by a surcharge applied to the cost of all water delivered for civilian consumption.

In the late 1950's, alternate ways to provide more water for the residents had been studied. These included a new aqueduct, small desalters along the Keys, a combination nuclear power/desalter, and hauling water by tanker. None was financially feasible unless the price of water was increased.

In 1964, the Commission engaged Fluor Corp. for a feasibility study of supplementing the aqueduct with a single-purpose desalting plant. It would be located in the lower Keys and would satisfy projected water needs from 1967 through 1982. Essential ground rules for establishing feasibility where: (1) that a new source supply additional water with no increase in the existing rate, and (2) that financial support could be obtained through sale of bonds at 4.5% interest and 35-year maturity. (The Housing and Urban Development Agency provided the financial stability by guaranteeing a 4%, 40-year. loan.)

The study indicated feasibility of installing two 2.62-mgd plants on Stock Island. The first plant was to operate in 1967 and the second in 1972. Possible water costs indicated by the study were around 90¢/1000 gal—a favorable figure.

In 1965, the Commission engaged Fluor as architect-engineer to develop preliminary designs, establish performance parameters and guarantees, and prepare construction and material specifications for the first unit. Westinghouse won the contract for design, procurement, construction, startup and testing under a total responsibility contract with the customer. Job scope included the complete facility seawater supply, brine outfall, boiler plant, fuel oil unloading and storage, desalting plant, buildings, product post-treatment and a two-mile pipeline to customer's system.

Upon plant completion, the Commission executed a contract with Westinghouse for management and operation of the plant. This continuity of responsibility has had definite advantages during analysis of problems, development of solutions, and for optimizing over-all plant operation.

Desalting plant design

The Key West facility employs a multistage flash evaporation process with brine circulation. It is a high-temperature design with maximum brine temperature of 250 F, and last stage temperature of 100 F. Feed at 78 F is taken from seawater wells adjacent to the evaporator. Feed treatment consists of acid addition with pH control, and two stages of degasification. All major pumps except the seawater pumps are provided with 100% standby and are turbine driven. Exhaust steam is piped to the brine heater to provide the evaporator heat source. The selection of turbine drives was basically one of economics because of the relatively high (17 mills/kwh) electric power costs.

Figure 1 Key West desalting plant. A single-unit, single-purpose plant with 2.62 mgd capacity.

The multistage flash evaporator has 50 stages—48 stages of heat recovery and two stages of heat reject. There is also a distiller cooling section wherein the distillate is cooled from a last stage temperature of about 98 F to a final temperature of 88 F. The evaporator is 105 ft long, 70 ft wide, 8 ft high. When flooded for hydrostatic testing, it weighs 2500 tons.

A central control room is located in the plant operations and maintenance building, adjacent to the evaporator. Normal plant operation and control is monitored from this control room. The plant is operated by two men per shift: the lead operator, normally stationed in the control room, and a roving auxiliary operator to tend the boiler and other plant equipment. Total staff numbers 14.

Field fabrication

This evaporator was built on a strip of fill land having very poor load-bearing characteristics. To provide adequate support for the equipment, over 220 reinforced concrete piles were driven to refusal—160 of them for support of the evaporator vessel alone. Steel plate was shipped to the site by barge for fit-up and welding in the field. Tube sheets and tube supports were machined prior to shipment. The 18-gauge, 70/30 copper-nickel tubes are 105 ft long. Shipping them from the tube supplier to the jobsite involved a special rail shipment with idler cars to Miami where they were transferred to a barge for shipment to Stock Island. There are more than 400 miles of tubing at Key West.

The evaporator itself is supported by a system of flex plates which are individually oriented. This allows for thermal expansion of the massive unit in several directions simultaneously, while the flex plates maintain their load-carrying ability. Temperature of the first stage flash chamber (at the lower corner) is about 246 F, while the opposite upper corner is at the temperature of incoming seawater, or 78 F. The entire vessel must be allowed to grow from its cold dimensions as it heats to these operating temperatures without undue stress on the many full-vessel-length field welds. This unique design task was accomplished through detailed stress analysis of the entire structure with advanced computer analysis techniques.

Steam for the recycle pump turbines and brine heater comes from a water tube oil-fired package boiler. The boiler has twin burners with duplex-type pump and heater set. To increase reliability, the boiler was built for a maximum continuous rating of 85,000 lb of steam per hour to satisfy a normal evaporator steam flow of less than 70,000 lb/hr.

Startup

Not unexpectedly, startup problems were encountered. In the early stages of operation, it was found that both the quantity and pressure of aqueduct water available for cooling and emergency boiler makeup were extremely low.

This reflected the existing inadequacies of the local water facilities. During surge conditions, it was almost impossible to maintain minimum safe water level in the boiler. Addition of a service water pump and pushbutton control at the control panel enable the operator to get an adequate supply of service water at high pressure whenever he needs it.

Based on experience of the local utility, City Electric System of Key West, original specifications required that the seawater supply be drawn from seawater wells. Although there are some disadvantages to this, it was felt that the advantages would far outweigh drawbacks. Wells do provide a clean filtered flow of seawater, free from the usual problems with marine life, sand, shells and algae. On the other hand, water taken from wells was high in hydrogen sulfide content—5 to 10 ppm. It was expected that this would reduce to 1 or 2 ppm over a period of months. Such had been the experience of City Electric System.

However, after several months of pumping, the hydrogen sulfide content remains at 5 to 6 ppm. First indications of the severity of hydrogen sulfide attack on monel alloy came after 30 days of operation; it was necessary to replace all of the mesh in the last stage of the evaporator. Shortly thereafter, the mesh in the 49th stage also had to be replaced. Since the last stage, and under certain conditions the 49th stage, are exposed to corrosive noncondensable gases, including the low residual hydrogen sulfide, mesh replacement was with 316 stainless steel.

Successful large-scale desalting

From the first day of operation, strainers at the suctions of the brine extraction and distillate pumps kept plugging up. This is a normal occurrence which should diminish shortly as residual construction debris is flushed from the system. However, the problem persisted with an unusual amount of mill scale showing up in the strainers. This was attributed to outside storage of steel plate at the site. To minimize outage time, the strainers were modified to permit cleaning without a shutdown. Cleaning strainers during full-load operation was routine through those early months.

Operating the desalter

Plant instrumentation and control equipment was designed for operation under all load conditions with a minimum of operating personnel—an operator and an assistant operator on each shift. All flows, temperatures, and pressures critical to plant operation are monitored, recorded and automatically or manually controlled at the evaporator control panel. In addition to operators, two men are assigned to equipment and instrumentation maintenance,

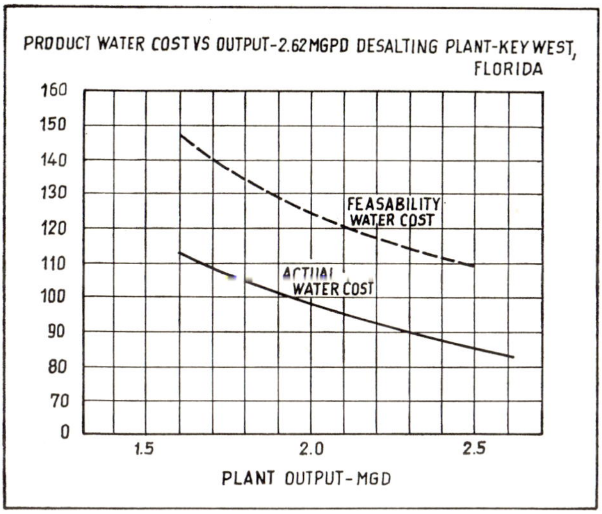
PRODUCT WATER COST VS OUTPUT-2.62MGPD DESALTING PLANT-KEY WEST, FLORIDA
160
150
140
130
120
110
100
90
80
70
0
1.5
2.0
2.5
PLANT OUTPUT-MGD
FEASABILITY WATER COST
ACTUAL WATER COST

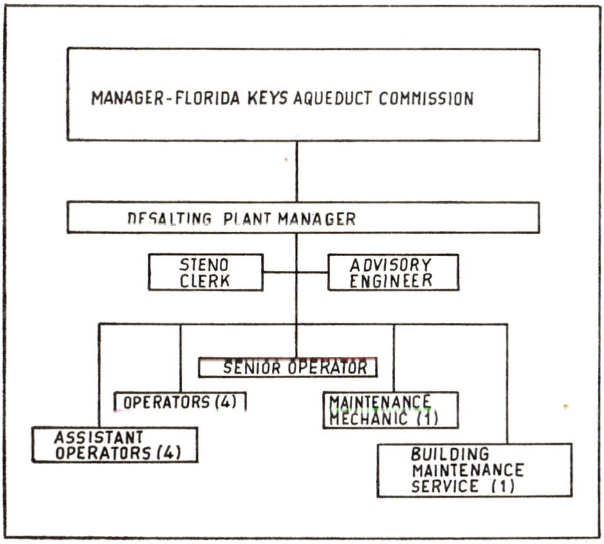
MANAGER-FLORIDA KEYS AQUEDUCT COMMISSION
DESALTING PLANT MANAGER
STENO CLERK
ADVISORY ENGINEER
SENIOR OPERATOR
OPERATORS (4)
MAINTENANCE MECHANIC (1)
ASSISTANT OPERATORS (4)
BUILDING MAINTENANCE SERVICE (1)

Figure 2

and there is a secretary with purchasing, payroll, accounts payable, and other office responsibility. Including the plant manager and the operating superintendent, a total of 14 people operate this plant for 24 hours a day seven days a week.

Acceptance tests took place in September of 1967. On September 4, 1967, the plant was taken to full-load operation in preparation for a final acceptance test. This officially began September 8, and the plant was operated continuously for 30 days with no equipment failure or malfunction.

Product water was pumped to storage from September 4 to October 16, a total operating period of more than 42 days. During this time, there were only three minor interruptions in the delivery of product water, each due to power interruptions which were beyond the plant's control. This acceptance run established the plant's capability for sustained operation, stability over long periods, and ability to respond rapidly to load change. As a result of the power interruptions, the plant showed it could be restored to service quickly.

Effects of feedwater

According to the design specifications, bidders were to assume feedwater with total alkalinity as calcium carbonate of 117 ppm. Well water alkalinity was considerably higher than anticipated, and as set forth in the specifications. This was verified by analyses and again by the rate of acid consumption.

According to specifications, the acid consumption guaranteed would have to be adjusted to compensate for this higher well water alkalinity. A curve indicating the required adjustment was included in the test procedures, and it was understood that well water alkalinity would become a critical part of the test data; therefore, it would have to be monitored accurately throughout the tests. Daily records show well water alkalinity remained at the 197-ppm level, while raw seawater in safe harbor continued at 121 ppm.

After the four-day performance test, equipment and piping were inspected to verify that the plant was in condition for continued operation. Inspection was made of the tubes, tube sheets, water boxes at the inlet to the heat reject condenser, the 48th stage condenser inlet, and the outlet of the first stage. In all cases, tubes and tube sheets were in good condition.

A thin black film was found in the tubes and on the tube sheets at the inlet end of the heat reject condenser. Later analysis showed this film to be a result of high H_2S level in the well water. Several pieces of coral rock were also found in the inlet water box. The shell side of the tubes was inspected in the first stage condenser, the 32nd stage condenser, and the distillate cooling condenser. The first stage tubes had a thin black film which was water soluble. The 32nd stage tubes had an even thinner film. The distillate cooler condenser tubes had a somewhat heavier film which could possibly affect heat transfer rates, consequently, the distillate temperature.

During the four-day performance test, average distillate temperature was 92.5 F, 4.5 degrees higher than the specified limit of 88 F. Inspection of heat transfer surface in the distillate cooler section and analysis of the deposits found showed that distillate temperature was indirectly affected by the high H_2S content of the well water. A chemical cleaning procedure was developed, using inhibited hydrochloric acid.

The plant was brought up to full capacity operation on November 19, and test data confirmed that the chemical cleaning had solved the problem of

Cost Data for 2.72-mgd Desalting Plant at Key West

	Total	¢/1000 gal[‡]
Plant Capital Cost:		
Construction cost—		
Turnkey desalting plant	$3,369,400	
5.0-million-gal product storage tank	218,000*	
Total construction cost	3,587,400	
Additional plant costs—		
Land	62,500	
Interest during construction	71,800	
Financing charges	45,000	
Architect-engineer services	290,000	
Total plant capital cost	$4,056,700	
Annual Cost:[†]		
Fuel oil	$261,990	27.8
Power	24,680	2.6
Acid	82,000	8.7
Caustic, anti-foam, boiler chemicals, limestone, lubricants, etc.	20,910	2.2
Maintenance materials, insurance	25,000	2.6
Administration, supervision, labor	170,180	18.1
Bond service and insurance	222,185	22.5
Total	$806,945	84.5

* Product storage tank was supplied by others and was not part of the plant contract. It should be included, however, for analysis of the facility economics.

† Derived from cost records for first year of operation.

‡ Based on production of 943,160,000 gal fresh water.

distillate temperature. At gross output of 2.72 mgd, distillate temperature was 88 F with seawater inlet temperature at 78 F.

The first stage flash chamber, mesh, and several of the lower stages (46-50) were inspected. The first stage distillate through had a black, flaking corrosion product on the steel plate opposite the tubes and adjacent to the mesh. The mesh was somewhat blackened, but otherwise in good condition. In all of the lower stages (46-49), flash chambers and mesh were in exceptionally good condition. There were some deposits on the walls below the mesh.

The 50th stage showed average corrosion for a deaeration stage; the corrosion appeared generally throughout the stage, but concentrated toward the low-temperature end of the stage. This is normal as gases and noncondensables are released in the degasifier and swept across the stage toward the cool end. Epoxy coating in the last stage flash chamber showed no evidence of flaking or exfoliation.

The boiler mud drum contained a normal accumulation, and all boiler tubes were found in excellent condition. Carbon deposits had collected on the furnace wall tubes in the area adjacent to the flame. This was attributed to fuel oil quality, and apparently it had no effect on boiler performance. All in all, plant inspection showed no more than normal rates of corrosion, and, in most cases, less than expected. There were no signs of undue stress or wear in machinery. These inspection results from Key West are further verification that materials and designs for large desalting plants are well in hand.

During the four-day performance test, the plant surpassed all guarantees concerning capacity and major consumables. Consumption of fuel oil, power and acid were respectively 2.85%, 6.06% and 2.12% better than guarantee. The uninterrupted test run of 42 days exceeded the specified test period by eight days, and the plant's stability enabled test data to be taken at fixed operating conditions over long periods.

Reference

1. Originally appeared in *Power Engineering*, July, 1969. Reprinted by Water Province Department, Westinghouse Electric Corporation, Philadelphia. The reprint is the article in its entirety.

V

A DESERT SEACOAST PROJECT AND ITS FUTURE[1]

Carl N. Hodges

AS INDICATED IN THE section of this book by Peveril Meigs, and elsewhere, about 20,000 miles of coastal desert is potentially available for human habitation, for the most part. Further, it is well demonstrated that the coastal desert areas provide one of the most desirable regions for human habitation, if the basic amenities of life can be supplied. Ample evidence of this fact is provided at the desert coasts of southern California (U.S.A.) and Baja California (Mexico) to which people are migrating rapidly. For the case of southern California, to supply the water necessary for this influx, aqueducts have already been built 340 miles long to the Colorado River. Other aqueducts over 400 miles long are under construction to the Feather River[1], and it has even been proposed that some day water may be brought from Alaska, thousands of miles away[2].

Recently, however, there has been growing interest in the world in the possibility of the utilization of desalted seawater as a resource alternative to importing fresh water from distant sources for use at coastal locations. Probably the most ambitious investigation into this possibility has been conducted by the Oak Ridge National Laboratories into the feasibility of utilizing large nuclear energy centers to support industrial and agro-industrial complexes[3]. Their study evaluated the potentialities of a central facility that could produce 2000 megawatts of electricity and a billion gallons per day of distilled water from the sea. This quantity of water would be adequate to support an industrial complex as well as a 300,000-acre farm.

The authors of the Oak Ridge report are optimistic that such a facility could be constructed in perhaps 10 to 30 years and be economically sound. They do point out, however, the dangers of projecting economic analyses that far into the future; they suggest that their report be taken strictly as a preliminary feasibility report and that certainly any actual construction in the future would require a detailed analysis for a specific site and location.

If such a facility were ever to be constructed, due to the size it undoubtedly would have to be done on a national or possibly international scale. Further, the capital investment would be so great that few developing countries of the world (which contain much of the coastal desert areas) could afford such a venture without great assistance from the developed world.

In 1961, the University of Arizona began an investigation into the possibility of utilizing solar energy as a thermal source to provide a means of economically supplying desalted seawater for relatively small coastal desert communities. Desalting may be a more attractive alternative to importation for small communities than for larger communities. This situation comes about because long aqueducts are generally justified only if relatively large quantities of water are required.

With the support of the Office of Saline Water, U.S. Department of Interior, in 1963, a pilot facility was constructed in cooperation with the University of Sonora[4] at Puerto Peñasco, Sonora, Mexico. This facility is located approximately 220 miles southwest of Tucson, Arizona, on the Gulf of California. The plant was operated for a period of two years, utilizing solar energy as the primary energy source. During this time the nature of the research program began to change.

As potential sites for future solar distillation plants were investigated, it became obvious that in almost all coastal communities of the world (and probably in all that could ever support any significant desalting plant) there would be available some waste thermal energy from either a municipal power center or individual private power-generation facilities. In general, the waste thermal energy would be more economical than solar energy when the capital investment necessary to construct the solar collector devices was legitimately amortized. Therefore, the Puerto Peñasco desalting plant was modified to operate from the waste thermal energy of the diesel-electric plant that generates the power for the experimental complex, and research on solar collectors was discontinued.

As further investigations were conducted into the actual requirements of coastal desert areas and the economics of desalting seawater, it also became clear that for both sociological and economic reasons it would be desirable to produce at least part of the food consumption of the community within the community. It was also evident that for anything but gigantic desalting plants, it is unlikely that in the foreseeable future desalted seawater would be inexpensive enough for conventional-type agriculture. Even for high-value items such as vegetables, the production of crops with irrigation water that costs $0.50 to $3.00 per thousand gallons would not be feasible.*

---

* The real cost of water from a desalting plant is very difficult to specify; however, most efficient plants now operating or projected for the near future produce water within this range.

A Desert Seacoast Project

A plant growing in an open-field condition will require water amounting to 100 to 1000 times its final weight. Most of this water serves no important physiological function except the cooling of the plant by means of transpiration. If the plant could be encased in a very humid environment in which the potential for transpiration could be greatly reduced, and at the same time a mechanism besides transpiration substituted for the cooling requirements of the plant, it is possible that plants could be grown on one to ten per cent of the fresh water that they would require in the open field. With a

Figure 1 University of Arizona-University of Sonora experimental facility, Puerto Peñasco, Sonora, Mexico.

relatively small requirement for water, the cost of the water would become much less important in the total crop economics, and food production could be accomplished in connection with present-day desalting technology. In an attempt to do this, with financial support from the Rockefeller Foundation, a system of controlled-environment structures for plant growth was added to the ongoing research program at Puerto Peñasco. They completed the central facility as envisioned for the production of power, water, and food for coastal desert areas (Figs. 1 and 2).

Power production

Power for the Puerto Peñasco experimental facility is generated in a manner similar to that utilized by many small communities of the world. A Caterpillar D330T, 80-horsepower, diesel engine turns a 60-kilowatt electric generator. The 60 kilowatts of electrical power is used for pumps and blowers in the desalting plant, for pumps and fans in the greenhouses, and for lighting and powering the experimental equipment throughout the facility. In any internal combustion engine, such as the diesel engine, approximately two-thirds of the fuel energy supplied to the engine is normally rejected as waste heat. Half of this waste heat is usually thrown away in the hot exhaust gases, and half is rejected through the water cooling system of the engine. However, the Puerto Peñasco engine is equipped with waste heat recovery equipment which is cooled by the salt water feed to the desalting plant. The waste heat from the engine warms the salt water to 160°F.

Water production

The Puerto Peñasco desalination plant is a 6000-gallon-per-day humidification type distillation unit. The humidification process was developed at the University of Arizona particularly for use with low-temperature energy sources. By operating the plant at temperatures below 160°F, no scaling from the seawater occurs and no pretreatment of the seawater is necessary.

The 160°F stream of seawater supplied from the power facility is pumped to a packed column evaporator (the round tower in Figure 2) where it flows countercurrent to a high-velocity stream of air. This stream of air is transported to an extended surface condenser (the square tower in Figure 2) which is cooled by 78°F seawater taken from a salt-water well. It is the water from the condenser (*preheated* by the latent heat of condensation) that is fed to the power plant where it is heated to 160°F. The water vapor in the airstream condenses on the fins of the surface condenser and falls to the bottom of the condenser, where it is collected to be pumped to storage.

Only 10 per cent of the salt water that is fed to the desalting plant is actually

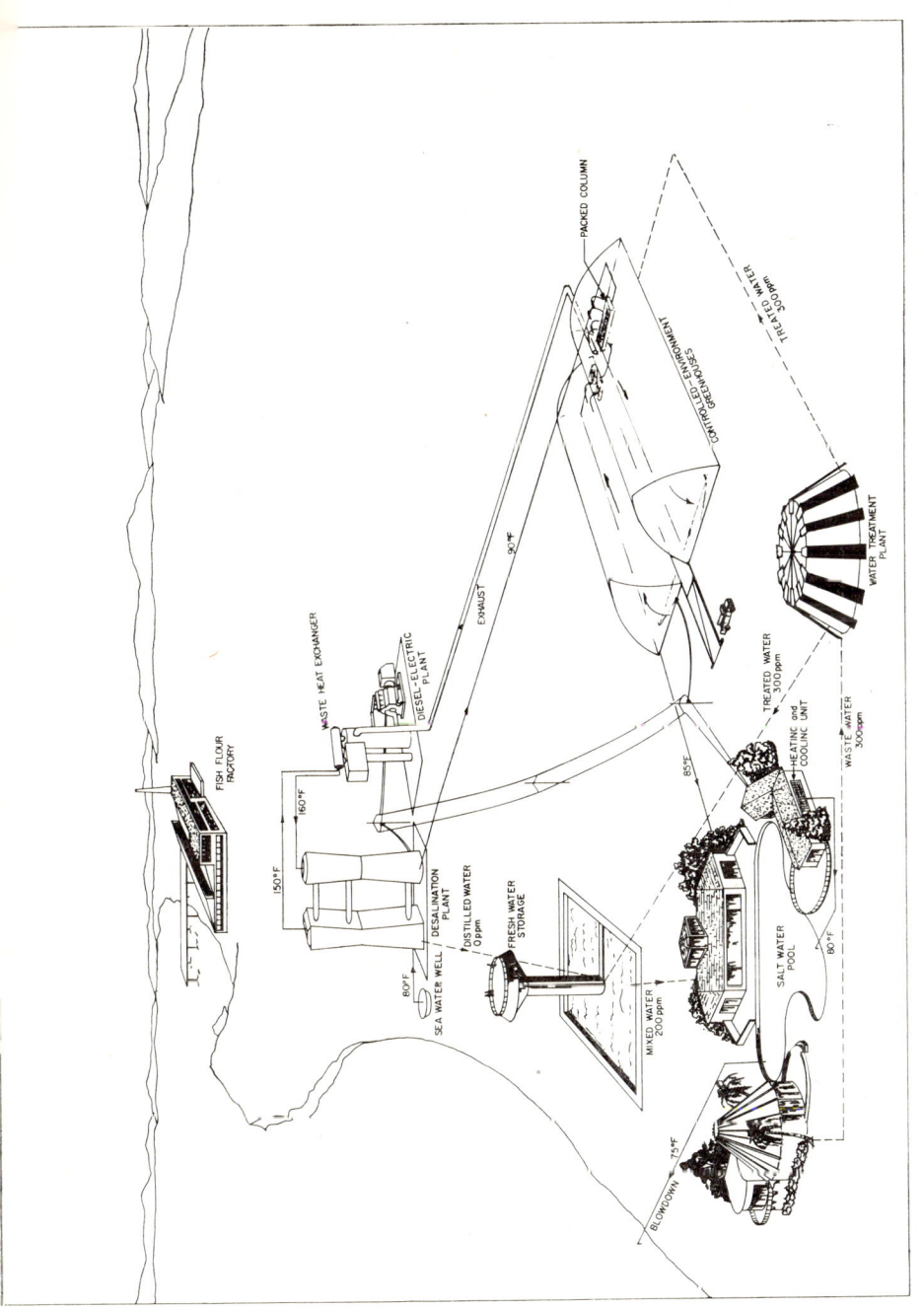

Figure 2 Schematic of the controlled-environment community shown in Figure 1.

evaporated and collected as product. The other 90 per cent of the seawater is rejected from the plant at approximately 88°F, still containing all the original salts. The 6000-gallon-per-day productivity of the pilot plant is more than adequate to meet the requirements of the 20,000 square feet of experimental greenhouse area and supply the drinking-water bottling facility operated by the University of Sonora for the residents of the city.

Food production

As shown in Figure 1, there are four experimental inflated greenhouses at the Puerto Peñasco facility. Each of the four greenhouses consists of two chambers connected by underground tunnels. The tunnel at one end serves as the entryway into the chambers as well as an air passage between the two halves. The tunnel at the opposite end contains a packed column heat exchanger for environmental control. The waste salt water from the desalting plant, as well as additional seawater, is continuously pumped into the greenhouse heat exchangers. The air within the greenhouse is circulated continuously countercurrent through the spray of seawater, so that by intimate contact with the seawater, the air is maintained at close to 100 per cent relative humidity and at a temperature very near to that of the seawater feed.

During the winter some of the seawater, which is continuously evaporating into the airstream, hic is condensed on the inside of the inflated plastic film. This condensation rains down inside the chambers to irrigate the plants. The amount of condensation is far greater than the amount of water consumed by the plants in the humid environment; in fact, during the winter months there is a net production of fresh water from the greenhouses; so it is not necessary to supply the agricultural system with any of the expensive distilled water from the desalination plant.

If crops are grown in the middle of the summer, however, some water from the desalination plant must be supplied to the greenhouses. This is necessary since the condensation within the house is not adequate during the hot weather to supply the requirements of the plants, and also during the warmest part of the day some condensation of fresh water (transpired from the plants and evaporated from the ground) occurs on the saltwater stream passing through the packed column.

For many installations it is possible that the growing of crops will not be desirable during the two warmest months of the year. Since the sizing of the mechanical equipment (pumps, fans, packed columns) is almost completely a function of the maximum temperature allowed in the house, it may be more economical to simply set aside the two warmest months of the year for renovation of the greenhouses. If an expendable plastic is utilized, as is now envisioned, these two nongrowing months may be used for such things as removing and discarding the plastic film, cleaning the greenhouses, renovating cultural systems, installing new plastic film, sterilizing, and replanting. If permanent plastic film is utilized, then possibly larger mechanical equipment for the additional cooling load will be justified, or a shading program initiated, to allow year around food production.

It has been determined that for some crops grown in these chambers, where many of the environmental factors can be controlled, the limiting factor on growth is the carbon dioxide level of the atmosphere surrounding the plants. By elevating the carbon dioxide level[5], significant increases in productivity can be obtained.*

Since the engine that provides power for the facility burns a hydrocarbon, one obvious source of carbon dioxide for the chambers is the exhaust gases. As shown in Figure 2, these gases are purified in a seawater scrubber and the carbon dioxide fed to the interior of the chamber in controlled amounts. Figure 3 is a photograph taken inside of the chambers, showing some of the vegetable crops that have been produced.

Considerable research and development work remains to be done on the total system; nevertheless, it can be said that the technical feasibility of this type of integrated facility has been demonstrated, and preliminary designs and cost estimates can be prepared for operational-scale installations. Figure 4 is an artist's concept of one such unit that is now being evaluated.

The facility consists of a central power plant composed of five diesel or natural gas turbines. Turbines have been selected for this hypothetical installation, since it is assumed to be located in a Middle East area with low fossil-fuel costs. Present industrial turbines are relatively inefficient, but their high heat-rejection rate has some advantages when the total energy requirements of the power, water, food system are considered. The desalting plant is a two-module, 64,000-gallon-per-day humidification unit.

The inflated greenhouses are similar to the ones now in operation in Puerto Peñasco, although larger and with one central service tunnel. The anticipated productivity from the greenhouses is over a thousand tons per year of high-quality vegetables. Such an installation then not only has the poten-

---

* The most responsive plant evaluated to date is IR-8 rice, which showed a 100 per cent increase when the carbon dioxide level was increased from 300 ppm, the normal atmospheric level, to 1200 ppm.

tial to supply the local community, but it might also export to interior regions.

To specify productivity costs from such a unit is difficult, due to the extreme effect on the cost of the amortization schedule selected, labor costs used, and miscellaneous other costs. One specific analysis was made for Puerto Peñasco, Sonora, Mexico, however, in which a 15-year amortization schedule for permanent equipment was utilized and actual local labor costs were considered; the projected production costs of power, water, and vegetables would then be slightly less than 50 per cent of the cost of present alternative sources.

It is the opinion of the author that systems such as that of Figure 4 will be utilized in some coastal areas of the world within the next few years. Undoubtedly such installations will be limited for some time to specific areas where economic factors most favor the expenditure of the capital necessary to construct such a facility. It is encouraging to note, although not likely to happen, that if it were necessary, with the development of 5 per cent of the

Figure 3 Vegetable crops inside Puerto Peñasco greenhouses.

Figure 4 Proposed power water-food facility for coastal desert areas.

world's desert coasts (925 miles) to a depth of 20 miles using such a system producing food at productivity rates already obtained, one billion people could be fed.

References

1a. Reprinted from an article in *Arid Lands in Perspective*, by permission of University of Arizona Press, Tucson, 1969. Reprint is article in its entirety.
1. MONROE, L. E. (ed.) Metropolitan Water District of Southern California report for the fiscal year July 1, 1966, to June 30 (1967).
2. NORTH AMERICAN WATER AND POWER ALLIANCE (NAWAPA) Brochure 606-2934-19. The Ralph M. Parsons Company, Los Angeles (1967).
3. OAK RIDGE NATIONAL LABORATORY REPORT. Nuclear Energy Centers, Industrial and Agro-Industrial Complexes, Summary Report. ORNL-4291. 30 pp. (1968).
4. HODGES, C. N., *et al*. Solar distillation utilizing multiple-effect humidification. Department of the interior. Office of Saline Water Research and Development, Report 194 (1966).
5. HODGES, C. N. Controlled-environment agriculture for coastal desert areas. University of Arizona, Institute of Atmospheric Physics, Environmental Research Laboratory, Report. 32 pp. (1967).

VI

1 ENGINEERING FEASIBILITY AND ECONOMIC STUDY FOR DUAL-PURPOSE ELECTRIC POWER-WATER DESALTING PLANT FOR ISRAEL[1]

Introduction

ISRAEL ANTICIPATES A serious water problem by the early 1970's. Action toward solution of this problem was taken when Prime Minister Eshkol and President Johnson signed a Joint Communique on June 2, 1964 calling for cooperation between the United States and Israel in the preparation of a study to determine the feasibility of a large desalting plant to be built in Israel. In October 1964, a Joint Team of experts from Israel (Water Planning for Israel, Ltd. (Tahal); Israel Electric Corporation, Ltd.; Israel Atomic Energy Commission) and from the United States (Department of Interior Water and Power Development Office and Office of Saline Water; and the Atomic Energy Commission) completed a preliminary study on a combination seawater desalting and electric power plant to be built in Israel. The study indicated that a plant of this kind may be technically and economically feasible in Israel. Based on these conclusions, a U.S.–Israel Joint Board recommended Kaiser Engineers and its principal subcontractor, Catalytic Construction Company, for the preparation of a detailed study of the above subject. In December 1964 the U.S. Department of Interior, with the concurrence of the U.S. Atomic Energy Commission and the State of Israel, contracted with the Kaiser/Catalytic team to conduct an engineering feasibility and economic study of a dual-purpose power generation and desalting plant to be located in Israel. The bases for the study were specified as:

1. A plant capacity of 175–200 megawatts of salable electricity and 100 to 150 million cubic meters of water per year.
2. A plant ready for initial operation in 1971 and commercial production in 1972.

The study was conducted in two phases.

The first phase, completed in July 1965, was directed toward determining the feasibility of a dual-purpose electric power-water desalting plant ready for commercial operation in Israel by mid-1972 and included:

1. Comparisons of various dual-purpose nuclear power and desalting plants using the multi-stage flash evaporation process and selection of one of these plants (hereinafter referred to as the reference plant) for detailed evaluation.
2. Comparison of this selected plant with a comparable fossil-fuel dual-purpose plant.
3. A recommended development program to confirm certain desalting plant components and design criteria.

The second phase was directed toward preparing a detailed evaluation of the reference plant and to develop information for use in preparing an application to funding agencies. During Phase II the conceptual design was refined and new estimates of the capital cost, annual cost, and the unit cost of water were prepared. In addition, several plant alternatives and factors important in the determination of the total annual cost were examined such as:

1. Differences in the desalting plant design and in the cost of water resulting from the selection of either a boiling-water or pressurized-water reactor for the nuclear steam supply.
2. Differences in the cost of water resulting from the selection of a concrete instead of a steel reactor containment building.
3. Differences in the cost of water resulting from the selection of concrete instead of steel evaporator chambers.
4. Review of the plant operating factor.
5. Determination of the effect of seasonal seawater temperature on the capital cost and the unit cost of water.
6. Determination of the change in the cost of water which would result if potential decreases in the nuclear fuel cycle costs materialize.

The refinement of the estimates of capital cost incorporates the information from preliminary quotations received from vendors of reactors, turbine-generator units, and complete desalting plants, and also of desalting plant components including brine heaters, recovery and reject stage tube bundles, pumps and vacuum system.

This report includes the results of the Phase I work and describes the results of the above Phase II studies the estimates of capital cost, annual cost, and the unit cost of water.

Summary

The results of the Phase I and Phase II work show that a dual-purpose power-water desalting plant in Israel with capacity of 200 megawatts (MWe) salable power and 100 million gallons water per day (MGD) of desalted seawater ready for initial operation in 1971 and full commercial operation by late 1972 is technically feasible, provided, however, that a timely development program such as is recommended herein is undertaken. This schedule is based on starting preliminary design of the plant in May 1966 and starting operation of the development program test modules by January 1968.

The dual-purpose plant upon which the cost estimates reported herein are based (reference plant) consists of a light water nuclear reactor supplying steam to a turbine-generator producing electric power; the 25 psia turbine exhaust steam in turn supplies heat to the brine heaters of the desalting plant wherein this exhaust steam is condensed before returning to the reactor. In the brine heater the turbine exhaust steam furnishes the thermal energy required by the desalting plant for evaporating fresh water from seawater. The concentrated brine resulting from the partial evaporation of seawater is returned to the sea.

Two available sites near Ashdod have been determined to be suitable for nuclear dual-purpose plants. Site 2 which is near Nitzanim is considered the preferred site; a subsurface investigation not yet completed at Site 2 will provide the required data to determine the actual plant location on this site and permit confirmation of the assumed subsurface structure design criteria. The frontispiece and Figure 1 (annotated) depict the layout of the reference plant.

Capital costs have been estimated for the dual-purpose plant. Separate cost studies were prepared by the Israel Electric Corporation and by Tahal to determine the cost of the electrical power and water conveyance facilities beyond the dual-purpose plant boundaries. Water production costs have been estimated for fixed charge rates of 5, 7 and 10 percent of the capital costs of the dual-purpose plant. The fore-going are based on 1965 costs. Unit water production cost estimates for the reference nuclear dual-purpose plant are as follows:

Engineering Feasibility for Desalting Plant for Israel

	Unit water costs at plant boundary		
Fixed charge rates	5%	7%	10%
Cents/1000 Gallons	28.6	43.4	67.0
Cents/Cubic meter	7.6	11.5	17.7
Agorot/Cubic meter	22.6	34.4	53.1

The importance of the fixed charge rate on the unit water cost can be noted from the above; a change of one percentage point yields a change of approximately 7 cents per thousand gallons.

Estimates were also made for fossil-fuel plants of comparable size. For all fixed charge rates through 10 percent, the unit cost of water from the reference nuclear dual-purpose plant was less than that from a fossil-fuel dual-purpose plant.

Table 1 Estimates of unit water cost at the plant boundary—1965 cost basis nuclear dual-purpose power-water desalting plant

NOMINAL FIXED CHARGE RATE	5%	7%	10%
CAPITAL COST*			
Depreciable Capital—Thousands of Dollars	$170,000	$180,200	$193,000
Nondepreciable Capital	17,000	17,000	17,000
Total Capital Cost	$187,000	$197,200	$210,000
ANNUAL COST: Thousands of Dollars			
Fixed Charges (Depreciable Capital)	$ 8,500	$ 12,610	$ 19,300
Fixed Charges (Nondepreciable Capital)	370	880	1,540
Operating and Maintenance Expenses	3,470	3,470	3,470
Fuel Expense	4,430	4,410	4,390
Total Annual Cost	$ 16,770	$ 21,370	$ 28,700
Power Credit	(7,900)	(7,900)	(7,900)
NET ANNUAL WATER COST	$ 8,870	$ 13,470	$ 20,800
WATER PRODUCTION			
Million Gallons per day	100	100	100
Million Gallons per year	31,050	31,050	31,050
Million Cubic meters per year	117.5	117.5	117.5
UNIT WATER COST			
Cents per Thousand Gallons	28.6	43.4	67.0
Cents per cubic meter	7.6	11.5	17.7
Agorot per cubic meter	22.6	34.4	53.1
Dollars per Acre Foot	93.1	141.0	218.0

* Excluding electrical transmission and water conveyance facilities.

48　　　　　　　　　　*Desalting Seawater*

Tables 1 and 2 on the following pages show details of the unit and capital costs for the reference plant.

To arrive at unit water costs, a determination was made of the value of the salable power. This value was credited against the total annual cost of the dual-purpose plant, and the difference, divided by the quantity of water produced, yields the unit cost of water. For this study, the Joint Board specified a power credit of 5.3 mills/kwhr, which is the cost of electric power produced in Israel at the prevailing fixed charge rates, and 85 percent plant operating factor, which value Kaiser Engineers has examined and agrees is reasonable for single purpose plants of 175 to 200 MWe salable capacity constructed in Israel in 1965. This value represents Israel Electric Corporation's estimate of the cost, on a 1965 basis, of producing power in a single-purpose (power only) plant designed to operate in the same capacity range as the dual-purpose plant studied herein; i.e., 175–200 megawatts electrical (MWe) salable power.

Table 2 Estimate of total capital cost—1965 cost basis nuclear dual-purpose power-water desalting plant (Thousands of Dollars).

NOMINAL FIXED CHARGE RATE DUALPURPOSE PLANT	5%	7%	10%
Depreciable Items			
Power Plant	$ 52,850	$ 52,850	$ 52,850
Desalting Plant	74,540	74,540	74,540
Intake and Outfall	4,950	4,950	4,950
General Plant Facilities	2,440	2,440	2,440
Other Construction Costs	16,980	26,090	37,560
Subtotal	$151,760	$160,870	$172,340
Contingency 12%	18,240	19,330	20,660
Total Depreciable Items	$170,000	$180,200	$193,000
Nondepreciable Items			
Land Cost	$ 0	$ 0	$ 0
Working Capital	16,000	16,000	16,000
Subtotal	$ 16,000	$ 16,000	$ 16,000
Contingency 6%	1,000	1,000	1,000
Total Nondepreciable Items	$ 17,000	$ 17,000	$ 17,000
TOTAL DUAL-PURPOSE PLANT	$187,000	$197,200	$210,000

Note It is estimated the additional investments required for electrical transmission facilities and water conveyance facilities are $5,000,000 and $25,000,000 respectively.

Engineering Feasibility for Desalting Plant for Israel

Based on the conceptual design work performed, it was concluded that either a pressurized water (PWR) or boiling water reactor (BWR) would be technically suitable for the nuclear steam supply. With respect to the reactors, the cost of a PWR would be essentially the same as a BWR; however, the equipment layouts are somewhat different. Typical layouts for both reactor types were prepared and are included in this report. Cost estimates for each were prepared and found to be competitive. The cost estimates appearing in this report are therefore equally applicable to the BWR or PWR reactor plant. The specifications for major equipment could be written on a "performance type" basis, in which event the manufacturers' competition between the alternates mentioned above would be preserved.

The estimates herein are based on a desalting plant of the multi-stage flash evaporation type consisting of four trains of 31 stages each. Each train would have four parallel module streams each with capacity of 6.25 million gallons of product water per day (MGD). The maximum brine temperature would be 220 F and the economy ratio would be 10.3 lb water per 1000 Btu heat.

The plant concept used as a basis for cost estimates includes engineering assumptions regarding process and equipment performance which represent an extrapolation of the desalting plant current state-of-the-art. Accordingly, there is recommended a development program which must be prosecuted in a timely fashion to make possible the scheduled start-up date, and the proper operating and cost performance of the plant.

Water and power background

The water and power needs of Israel are based on recent studies made by Tahal and Israel Electric Corporation. Based on these studies, Tahal forecasts that Israel's water consumption by 1970 will reach 90 per cent of the country's potential water resources that could be made available by that time. This critical situation demands that an additional source of fresh water be provided. The increasing salinity of the natural supply aggravates the problem of water shortage. One significant man-made water supply addition is the reclamation of sewage. But this aggravates the salinity problem still further because the effluent from this supply has a higher salinity than the original input water; thus, a supply of fresh water for dilution is mandatory. Even after the successful implementation of plans for conserving and reclaiming water, Israel will need to supplement its annual water supply by about 200 million cubic meters (MCM) by 1980, of which about 120 MCM will be required by the early 1970's. A desalting plant capacity of 100 million gallons per day will produce 118 MCM per year. In 1972, this quantity would represent approximately 8 percent of Israel's total demand. The provision of this high-purity supply will make it possible to reduce the salinity of water supplied to agriculture to the level considered safe for continued healthy crop growth; notably citrus which is a vital export.

Israel Electric's forecast of power demand and energy consumption is based on a detailed analysis of the expected consumption pattern of the main consumer groups in the period of 1970. The forecast concluded that by 1970 the "doubling period" of the maximum demand will be eight years, and the optimum unit size for additional generating units in the beginning of the next decade lies between 175 and 200 MWe bet.

State-of-the-art

Commercial desalting plants are operating successfully in many parts of the world. The process most advanced, and the one most used in commercial operation, is the multi-stage flash evaporation process; there are 10 MSF plants of at least one million-gallons-per-day capacity in operation; the largest unit has a capacity of approximately 1.5 million gallons per day. For scaling-up to the capacity proposed herein, the design of some parts of the process and some of the components must be confirmed in a development program to be discussed later in this summary.

Commercial production of electric power from a nuclear energy heat source is a proven operation in plants almost as large as that required for the reference dual-purpose plant. Other nuclear power plants, already under construction or on order by United States utilities under firm guarantees as to output and costs, are larger than that contemplated herein. These larger plants will be in commercial operation before Israel's dual-purpose plant would be completed. Both the pressurized water (PWR) and boiling water (BWR) nuclear reactor types, which have been considered equal alternates for the Israel plant, are being guaranteed in thermal ratings more than double that required for the Israel plant.

Reference dual-purpose plant

One site under consideration (Site 1) is 10 km north of Ashdod, and the other one (Site 2) is 9 km south of Ashdod, both being on the coast. Approximately 40 acres (160 dunams) are required for the proposed plant. Either of these locations will accommodate several plants of this size, thus ensuring

Engineering Feasibility for Desalting Plant for Israel

room for expansion. With respect to accessibility, subsoil conditions, topography, oceanography, transportation, communications, and availability of construction labor, it is concluded that both sites are equal. Particular attention was given to the nuclear hazards comparison between the two sites; it was concluded that both sites are acceptable, and that, depending on the direction of Ashdod's future expansion, there may be a slight preference for one site over the other. At present writing, the U.S.-Israel Joint Board has stated that Site 2 is preferred.

The conceptual design of the plant was based on certain ground rules, the principal ones being the following:

> 175–200 megawatts salable power credited @ 5.3 mills/kwhr.
> 100–150 million cubic meters/year water output.
> Fixed charge rates of 5, 7 and 10 percent.
> Estimates for reactor, turbine-generator and major desalting plant equipment of U.S. manufacture.
> Plant ready for commercial operation by mid-1972.

A survey was made of all contending U.S. reactor manufacturers to determine those which could supply reactors for commercial operation in 1971 for a plant of this size. Only the manufacturers of the PWR and BWR types indicated that their reactors would be available commercially, and were able to supply operating characteristics and cost data at this time.

Based on a computer optimization, a nuclear plant of about 1250 megawatts thermal (MWt) with a 250 megawatts electrical (250 MWe) (200 MWe salable) turbine-generator supplying steam at 25 psia to the desalting plant was chosen. Either a pressurized water (PWR) or boiling water reactor (BWR) would be suitable and the cost of water from a desalting plant which is optimized for one reactor type will be within one percent of the cost of water from a desalting plant which is optimized for the other reactor type. Heat balances and arrangement drawings for each are included in this report. In addition to the normal back-pressure turbine exhaust system, a provision is made in each cycle to supply additional steam to the brine heater through a turbine line with pressure-reducing station, thus ensuring continuous water plant operation when the turbine-generator is shut down and providing for maximum use of the desalting plant during the winter when colder sea temperatures will permit more water production. The pressurized or boiling water reactor systems are housed in containment or pressure-suppression type reactor buildings, respectively, which will contain the radioactive products released in the unlikely event of a rupture in the reactor coolant system.

The desalting plant conceptual design was based on the multi-stage flash evaporation process. This design was studied in sufficient detail to enable the determination of a development program to test those items of process and component design requiring validation in time to ensure commercial operation by the scheduled date. The plant on which cost estimates are based consists of 31 stages through which flow several streams: brine in tubes, brine in open channels, flashed vapor, and distillate (product water). Different temperatures and pressures exist in each stage. Because of the need to examine plants over a large range of physical and economic parameters, a computer program was devised to permit the evaluation of several thousand plants in a reasonable time, and to determine the optimum plant—the one with the combination of characteristics yielding the lowest unit water cost. The reference plant consists of four trains each having a capacity of 25 million gallons per day (MGD), and each capable of operating independently of each other. Within each train, there are four module streams operating in parallel, each having a capacity of 6.25 MGD and representing only a four-fold capacity scale-up of existing plants of this kind.

The reference plant cost estimate is based on field-erected steel evaporator structures. Should the development program confirm the suitability of the concrete structure alternate, a reduction in the capital cost of the dual-purpose plant of 3.6 million dollars would be realized. The major equipment (brine heaters, removable evaporator heat exchange bundles, pumps, air ejectors and gantry crane) is shop-fabricated with some field assembly to reduce freight costs and U.S. dollar expense.

Design is arranged so that if the turbine-generator of the power plant is shut down, the desalting plant can continue to operate; conversely, if the desalting plant must be shut down, the turbine-generator can continue to generate power. The optimization program yielded the following plant characteristics for the reference plant conceptual design:

Capacity	100 MGD
Trains	4
Stages	31
Maximum brine temperature	220 F
Economy ratio	10.3 lb water/1000 Btu heat
Steam pressure at brine heater	25 psia
Tube bundle material (recovery and reject stages)	7/8 in. diameter 20 BWG 90–10 copper-nickel alloy
Tube bundle material (brine heaters)	1 in. diameter 18 BWG 90–10 copper-nickel alloy

Tube bundle material (reject-deaerator stage and air cooling sections in all stages)	7/8 in. diameter 20 BWG stainless steel/90–10 copper-nickel bimetallic
Chlorine consumption	140 metric tons/year
Sulfuric acid consumption	33,000 metric tons/year

Preliminary quotations based on performance specifications were requested and obtained from several manufacturers of desalting plants and desalting plant components. The consensus of their responses is that the industry will be in a position to contract, in accordance with the schedule requirements of this project, to supply plants on an erected basis or plant components and equipment. The cost estimates for plants supplied on an erected basis are consistent with the cost estimates contained herein (which are based upon purchase of components and equipment).

After selection of the nuclear dual-purpose plant, a fossil-fuel dual-purpose plant was studied and compared with respect to characteristics and cost. The study showed that the cost of water produced in a nuclear dual-purpose plant was lower than that produced in a fossil-fuel dual-purpose plant for all fixed charge rates through 10 percent when the desalting plant capacity was 100 MGD.

Three important conclusions were derived from the optimization studies, and from the studies of reactor types:

1. The fixed charge rate has almost no effect on the selection of the technical specifications for a nuclear or fossil-fuel dual-purpose plant. The plant selected for the 7 percent fixed charge rate case will result in unit water costs, when evaluated at 5 percent and 10 percent fixed charges, no more than 1 percent higher than the water costs obtained from plants optimized for those fixed charge rates. It is therefore possible to proceed with one plant design even though the fixed charge rate may not be settled until a later date.

2. The value selected for power credit has no effect on the technical specifications for a dual-purpose plant of a given power and water capacity. The "optimum" plant will still have the same characteristics and lowest total annual cost when compared to others considered, with or without the power credit. The unit water cost will, of course, change with a change in power credit.

3. The desalting plant optimized for one reactor system was found to be essentially the same as that optimized for the other reactor system. Thus the same MWe rating would be required for either a PWR or BWR except for the small difference in nuclear reactor auxiliary power requirements and the

selection of the reactor type can be made as a result of competitive bidding with due allowances for normal nuclear power plant BWR-PWR differences and allowances associated with the handling of radioactive steam in the desalting plant brine heaters and steam driven auxiliaries.

Cost estimates

The capital cost estimates are based on 1965 costs and reflect information received by personal contact in Israel with construction contractors, by detailed discussions with Israel Electric and U.S. equipment manufacturers, and by review of the extensive experience of the firms making this study. It is believed that due allowance has been made for such things as maximum practical use of Israeli products and construction participation; availability in Israel of special skills required for a plant of this nature; and items inadvertently omitted at this conceptual design stage.

The capital cost estimates include, as separate line times, the costs of the electric transmission and water conveyance facilities required to bridge the gap from the dual-purpose plant boundaries to the main Israel distribution systems. These costs have not been reviewed during this study and are included as submitted by the U.S.-Israel Joint Board.

The unit water cost estimates are based on 1965 costs with respect to the capital costs upon which the fixed charges are calculated, and the costs for operating labor, materials and power credit. Fixed charges constitute the major portion of the total annual cost of the dual-purpose plant (Table 1); within the fixed charge rate, interest and sinking-fund depreciation (with its heavy dependence on interest rate) constitute almost the entire fixed charge. For instance, with a fixed charge rate of 7 percent, interest is 4.6 percent and depreciation (30-year sinking fund basis) is 1.6 percent—the two together constituting 90 percent of the total fixed charge.

The major systems of the reference plant described herein were analyzed as to component and combined availability for full-load operation over the life of the plant. It was concluded that the nuclear steam supply would be available to supply steam to the power plant and the desalting plant 90 percent of the time, the turbine-generator would be available 89 percent of the time, and the desalting plant would be producing 85 percent of the time. From an operational point of view, it is not possible to run the power plant at full load during the whole period of its availability and power production is therefore based on an 85 percent plant operating factor. The desalting plant has an availability of 85 percent and since no such restriction applies to water, the desalting plant operating factor is also 85 percent. These plant operating factors are average for the 30-year lifetime of the plant.

Another consideration important to the unit cost of water is the effect of the seasonal seawater temperature variations on plant capacity. The performance of the plant was determined for the various seawater temperatures that could be encountered. It was found that to maintain the average capacity of the desalting plant at the design value, it would be necessary to have sufficient reactor capacity to provide for operating the brine heater at the design temperatures and pressures at all seasons. This will require operating the nuclear steam supply at 1330 MWt instead of the normal 1250 MWt. The stretch capability of the nuclear steam supply can accommodate this increase. The cost estimates include the cost of the required reactor and feedwater heater capacity to maintain the average capacity at the desalting plant in spite of seasonal temperature variations.

The present estimates are believed to be quite conservative since they are based on vendors' quotations and on construction experience in Israel. It is prudent, however, to examine the sensitivity of the capital and unit water costs to the major design and cost parameters. For example, one leading reactor manufacturer has recently published predictions of fuel cycle costs for the period of the 1980's and 1990's which are 35 percent lower than those shown in Table 1. This decrease in fuel cycle cost would result in a decrease of 4 cents/1000 gallons in the reference plant unit water costs. The effect on the capital and unit water costs of such variants as lower fuel cost, longer and shorter plant life, and changing other major design parameters is reported in Section XI, Effect of Variants.

The effect on unit costs of the conservative value of 220 F for the maximum brine temperature was studied. Accordingly, the reference plant design was checked for its suitability for 235 F operation and was found adequate. Data now available on scaling properties of seawater at this elevated temperature also seem to be favorable as long as the blowdown concentration ratio does not exceed 2.0. If one optimized the reference plant using a maximum brine temperature of 235 F instead of 220 F, there would be some reduction in the cost of water.

Schedule

In attaining the late 1972 commercial operation date, the dual-purpose plant construction schedule is dependent on these sequential events: start of dual-purpose plant preliminary design by May 1966; ordering the nuclear steam supply by May 1967; start operation of the test modules by January 1968; and "freezing" desalting plant design by June 1968.

It should also be noted that the schedule and cost estimate are based on a steel spherical containment building in the event there is a PWR nuclear steam supply. It is presently estimated that concrete containment design for a PWR would result in a delay of about six months in the completion of the plant. The schedule for the concrete containment may be subject to further slippage due to the critical scheduling of certain equipment items which must be installed at specific times during the concrete containment construction.

Recommended development program

One important basis for this study is that the reference dual-purpose plant will be an operating utility rather than a developmental type installation. Therefore either the process and component technology must be proven at the present time, or there must be reasonable assurance that whatever development work will be necessary to "prove out" those features not currently in commercial operation will be successfully concluded by the required date of June 1968 for "freezing" desalting plant design. The study has shown that the only items requiring development are part of the desalting plant. Such things as hydrodynamic and process design to obtain the vapor separation desired; and the acquisition and evaluation of oceanographic and chemical properties information on the Mediterranean seawater to be used in the selected plant comprise the principal items of the recommended development and investigative program.

The testing of a full scale module is now planned by the OSW for the San Diego Saline Water Test Facility. This will provide information on such aspects of flash evaporation design as:

1. Tube bundle design.
2. Evaporator hydrodynamics.
3. Large pumps for recycle service.
4. Special mechanical hardware.

The portion of the development program to be accomplished in Israel—the work on site oceanography and seawater chemical properties—appears to have sufficient time for orderly prosecution of the work.

Without the results of the development program the design of the reference plant would have to be more conservative than would otherwise be necessary, thereby resulting in significant capital and operating cost increases as well as longer start-up time.

Conclusion

On the basis of this study, it is concluded that a dual-purpose power-water desalting plant of 200 MWe and 100 MGD capacity for Israel is technically feasible for initial operation in 1971 and full-commercial operation by late 1972. The demand for power and low-salinity water requires a solution such as a dual-purpose plant. The conceptual design on which the cost estimates are based employs certain desalting plant process and component design criteria requiring confirmation in a development program. However, these design criteria are considered to be reasonable extrapolations from currently proven technology. It should be further noted that the scheduled 1971 plant operating date is dependent on (1) the expeditious start of the module testing program and its completion in accordance with the schedule set forth herein, and (2) the starting of preliminary design of the dual-purpose plant by May 1966.

The matter of the economic feasibility of the utilization of the quantity of water to be produced at the cost estimated for the different fixed charge rates will have to be determined.

PART 2 EFFECT ON COSTS OF INCREASING CAPACITY TO 300 MEGAWATTS[2]

Introduction

In January 1966 Kaiser Engineers, in association with Catalytic Construction Company, completed an engineering feasibility and economic study for a dual-purpose electric power-seawater desalting plant for Israel. The results of this study showed that it was technically feasible to have a plant producing 200 megawatts of salable electrical power (mwe) and 100 million gallons of water per day (mgd) in commercial operation by late 1972, provided that, among other things, preliminary design was started by May 1966. Preliminary design has not yet commenced and the earliest commercial operation date now possible is mid-1974. The U.S.-Israel Joint Board recognized that one implication of this delay was that the Israeli power network would be able to absorb a larger block of salable electricity, say—300 mwe, by the later commercial operation startup. Therefore, the Board requested Kaiser/Catalytic to explore the changes in plant economics resulting from 200 to 300 mwe. Water production was to be maintained at 100 million gallons per day.

The results of this study are summarized in the next section. Following the summary are sections which describe the following:

1. The methodology used in performing the study.
2. Characteristics of the new reference power plant together with a discussion of nuclear fuel costs and plant heat balance.
3. Characteristics of the new reference desalting plant including the computer print-outs of the reference desalting plant internal performance details.
4. Method used in updating 1965 costs to 1967.
5. Estimates of capital cost, annual cost and unit water cost, 1967 basis.
6. Construction schedule bar chart and listing of the sequence of major events.
7. Effect of Variants table.

Summary

The results of this study reveal that a dual-purpose electrical power-water desalting plant for Israel with a nominal capacity of 300 megawatts salable electric power (mwe) and 100 million gallons water per day (mgd) can achieve commercial operation as early as mid-1974, assuming that preliminary design is commenced early in 1968. The study further reveals that, for all fixed charge rates, the unit cost of water from the 300-mwe, 100-mgd is lower than the unit cost of water from the 200-mwe, 100-mgd plant previously studied. In determining the unit cost of water from the 300-mwe, 100-mgd plant, a power credit of 5.0 mills per kwh (estimated by the U.S.-Israel Joint Board as appropriate for a 300-mwe base load plant) was used. This compares to a power credit of 5.3 mills per kwh used in the previous study for a 200-mwe base loaded plant.

This study not only includes consideration of the increased electrical plant capacity, but includes updating of costs to a 1967 basis and considers the technological improvements which have occurred in nuclear reactors since the previous study was performed. One of the most important developments in nuclear power plants which has occurred since the 1965–66 study is the advent of standardized commercial nuclear power reactors. These reactors offer cost and delivery benefits over nonstandard sizes, therefore this study was limited to standard size reactors.

While cost escalation of the desalting plant has been accounted for in this study, technical improvements in desalting plant design since 1965 have not been included—a factor which provides conservatism to the economics of the desalting plant portion of the dual-purpose plant.

Utilizing a standard size Boiling Water Reactor (BWR) and a standard Pressurized Water Reactor (PWR), the reference plant characteristics (plant resulting in the lowest unit cost of water) were determined as shown in Table 3.

The capital, annual and unit water costs were determined for both the BWR and PWR plants. The unit water costs associated with a BWR were slightly lower than those associated with a PWR for the conditions studied the difference being primarily the slightly higher net salable power available from the BWR plant. For the purpose of this study—to determine the effect of an increased salable electrical power generating capacity and an extended plant startup date on the unit cost of water—only BWR capital, annual and unit water costs are reported herein. However, general conclusions from the results of this cost analysis apply to PWR as well as the BWR reactor con-

Table 3 Reference plant characteristics

Reactor type	BWR	PWR
Reactor thermal rating, mwt	1593	1650
Final feedwater temperature, F	390	427
Electrical generation (gross), mwe	351	349
Net salable power, mwe	298	290
Turbine exhaust pressure, psia	26	26
Water plant capacity, mgd	100	100
Number of trains	4	4
Module streams per train	4	4
Number of stages	19	19
Terminal temperature difference (ttd), F	4.4	5.2
Maximum brine temperature, F	220	220
Heat to brine heater, mBtu/hr	4120	4325
Economy ratio	8.43	8.03
Concentration ratio	2.0	2.0

cepts. Cycle improvements which were not considered (such as reheat) may reduce the cost associated with a PWR plant, *and no conclusions regarding a preference for either type reactor should be implied from this study.*

The capital, annual unit water costs were updated to a 1967 basis by applying factors developed from the Engineering News Record. The resulting capital and unit costs are:

300-mwe, 100-mgd Plant	Fixed Charge Rate		
	5%	7%	10%
Capital cost *(1967 basis)*	$212,700,000	$224,100,000	$238,400,000
Unit water cost (¢/1000 gal)	23.6	40.2	66.6

The costs reported in the 1965–66 study for a 200-mwe, 100-mgd dual-purpose electric power-water desalting plant, 1965 basis are:

200-mwe, 100-mgd Plant	Fixed Charge Rate		
	5%	7%	10%
Capital cost *(1965 basis)*	$187,000,000	$197,200,000	$210,000,000
Unit water cost (¢/1000 gal)	28.6	43.4	67.0

It should be noted that the 200-mwe, 100-mgd plant capital cost should be escalated to a 1967 basis to be comparable to the 300-mwe, 100-mgd capital cost reported above. Were this done, the unit water costs of the smaller plant would increase, indicating that the actual unit water cost benefits due to the greater capacity are more than the differences shown in the above tables.

In attaining the mid-1974 commercial operation date, the dual-purpose plant construction schedule is dependent on these sequential events:

Commence preliminary design	January 1968
Order nuclear steam supply and turbine-generator	January 1969
Order containment structure	July 1969
Order major desalting equipment	February 1970

Conclusion

On the basis of this study, it is concluded that a dual-purpose power-water desalting plant of 300-mwe and 100-mgd capacity for Israel is technically feasible for commercial operation by mid-1974, and will produce electrical power and desalted water at unit costs lower than the unit costs of power and water produced in the 200-mwe, 100-mgd plant described in the 1965–66 study.

References

1. Prepared for United States-Israel Joint Board, January 1966. Kaiser Engineers, Oakland, California, in association with Catalytic Construction Company, Philadelphia, Pennsylvania. Reprinted portion is the Introduction and Summary.
2. December 1967. (This report bears the same general title as above, and was prepared by the same organizations under the same auspices.) Reprinted portion is Introduction and Summary.

VII

CONSTRUCTION OF NUCLEAR DESALTING PLANTS IN THE MIDDLE EAST HEARINGS[1]

Thursday, October 19, 1967

UNITED STATES SENATE,

COMMITTEE ON FOREIGN RELATIONS,

Washington, D.C.

The committee met, pursuant to recess, at 10:05 a.m., in room 4221, New Senate Office Building, Senator John Sparkman, presiding.

Present: Senators Sparkman, Gore, Lausche, Hickenlooper, Aiken, Case, and Cooper.

Senator SPARKMAN. Let the committe come to order, please.

We meet today to take testimony on Senate Resolution 155. This resolution, which has been submitted by the junior Senator from Tennessee, Mr. Baker, calls for the construction of a number of nuclear desalting plants in the Middle East.

(The resolution follows:)

[S. Res. 155, 90th Cong., first sess.]

Resolution

Whereas the security and national interests of the United States require that there be a stable and durable peace in the Middle East; and

Whereas the greatest bar to a long-term settlement of the differences between the Arab and Israeli people is the chronic shortage of fresh water, useful work, and an adequate food supply; and

Whereas the United States now has available the technology and the resources to alleviate these shortages and to provide a base for peaceful cooperation between the countries involved: Now, therefore, be it

Resolved, That it is the sense of the Senate that the prompt design, construction, and operation of nuclear desalting plants will provide large quantities of fresh water to both Arab and Israeli territories and, thereby, will result in—

(1) new jobs for the many refugees;

(2) An enormous increase in the agricultural productivity of existing wastelands;

Construction of Nuclear Desalting Plants in the Middle East 63

(3) a broad base for cooperation between the Israeli and Arab Governments; and

(4) a further demonstration of the United States efforts to find peaceful solutions to areas of conflict; and be it further

Resolved, That the President is requested to pursue these objectives, as reflecting the sense of the Senate, within and outside the United Nations and with all nations similarly minded, as being in the highest national interest of the United States.

Senator SPARKMAN As I understand it, the plan embodies the proposal made by former President Dwight D. Eisenhower and Adm. Lewis L. Strauss, former Chairman of the Atomic Energy Commission. I am pleased to see that Admiral Strauss is here this morning.

The committee is pleased to welcome the Senator from Tennessee, Mr. Baker, who wishes to make a brief statement on his resolution.

Let me say we are grateful to Senator Baker for suggesting to the committee the list of witnesses that we are to have this morning.

Senator Baker, we are glad to have you.

Statement of Hon. Howard H. Baker, Jr., U.S. Senator from Tennessee

Senator BAKER. Thank you, Mr. Chairman.

Mr. Chairman, and members of the committee, with your permission, rather than read verbatim the statement which I prepared in this connection, I would like to submit it for the record.

Senator SPARKMAN. It will be printed in the record in full.

Senator BAKER. Then, if I may, I would like to comment for a moment or two about the general nature of this undertaking, the purposes which hopefully it may encompass, and the function of Senate Resolution 155.

(The full statement of Senator Baker follows:)

Statement of Senator Howard H. Baker, Jr., before the Foreign Relations Committee on Senate Resolution 155 (Project MEND), October 19, 1967

Water, work and food are inextricably interwoven with the problem of creating a climate of peace in the strife-torn Middle-East.

These are among the basic, underlying causes of dissension among the peoples of the Middle East, and until these needs are sufficiently satisfied, they will continue to be key points of friction.

But imagine, if you will, a Middle East where barren, brown desert has been partially transformed into fertile green fields sufficient to produce large quantities of food, moving closer to the Biblical description of a "land flowing with milk and honey".

Water is a basic and unreplaceable need. If water is available, both Arabs and Israelis can begin to cultivate the fields that now lie fallow and produce ample food for themselves.

This is not an idle dream. It is now entirely within our grasp.

Project transcends partisan consideration

In the middle 1950's. President Eisenhower excited an assembly of the United Nations when he outlined his now famous "Atoms For Peace" program. He dramatically explored the possibility of using nuclear energy to provide fresh water for the thirsty peoples of the Middle East.

The need for providing abundant water has been recognized by Presidents Truman, Eisenhower, Kennedy and Johnson. Senators Anderson and Javits, Secretary Udall, Chairman Seaborg, Commissioner Ramey, Admiral Strauss, Dr. Weinberg, and others have participated closely in the development of the concepts and technology that have produced our capability for making this vision come true.

Thus, this project embodied in S. Res. 155 transcends partisan consideration. Able men and strong personalities of all political persuasions have worked unselfishly toward developing the ideas involved here.

The obstacles to this project have not been lack of vision or dedication. Rather progress has been slowed by the requirement that there be a careful development of the technology that will make the project economically and technically feasible.

Because the renewal of hostilities this year coupled with the tremendous advances in nuclear technology during the past few years made a permanent solution to Middle East problems even more urgent and feasible, the plan has been revived and refined.

The plan involves an exciting concept, the use of nuclear energy to make peace instead of war, the unleashing of man's mightiest energies to create rather than to destroy.

Since the explosion of that first atom bomb in the New Mexican desert in 1945, the world has lived under the awesome shadow of nuclear annihilation. This shadow grows darker with each passing year as the destructive force of nuclear weapons becomes greater and as more and more nations develop nuclear capabilities.

A start has been made in finding ways to utilize the atom for peace. In this project, as elaborated by Admiral Strauss, we have a way to make a practical demonstration of how the atom not only can be used for a peaceful purpose but can actually be used as a catalyst for creating peace.

What the plan envisions

In general terms, the plan envisions the construction, first, of a 6600 megawatt thermal desalting reactor which would produce 450 million gallons of fresh water per day. Two other plants, one of equal size to the first and one slightly smaller, would also be constructed.

Each plant would produce a quantity of fresh water greater than the combined flow of the major tributaries of the Jordan River. This water would be used for irrigating fields and turning the desert into productive land for farming.

It also would provide, virtually as a by-product, tremendous quantities of electrical energy which would be an incentive for industrial development of the region.

The project would from its inception provide a source of employment for the thousands of unskilled refugees whose inability to find work exacerbates the tensions of the Middle East.

These unskilled refugees could be utilized in the construction of the plants, the transmission lines, the irrigation ditches, the water canals.

Later, they would find jobs in the industries that would locate there and on the newly created farm lands.

The construction of nuclear desalting plants is fully within our present technology and capability. I will leave a more detailed study of the technical aspects of the plants to those who are fully abreast of all of the details.

An unusual and unique method of financing is one of the features of the plan.

Comsat-type approach

Admiral Strauss has proposed that a corporation be formed along the lines of COMSAT. (I have suggested that such a corporation might be appropriately named MEND—Middle East Nuclear Desalting.)

We have full reason to believe that the international financial community will enthusiastically support such a financial arrangement for the project and will therefore provide the principal source of capital for it on a risk basis.

Governments, including our own, and individuals would be allowed to participate on the same basis, as investors in a financially sound undertaking.

The COMSAT-type approach offers some enticing advantages from the financial standpoint. However, the principal advantage, the one which makes this venture attractive, is one which has received only token attention so far:

That is that such a corporation would have no national identity; it would be international in scope and identity.

Communications stalemate in Middle East

The Middle East crisis obviously has reached a communications stalemate. The Arabs won't talk to the Israelis, and the Israelis won't talk to the Arabs. Other governments have tried without apparent success to bring the warring parties together. International organizations have met with a similar and notable lack of success in their efforts at bridging the gap.

The establishment of an international corporation such as MEND, which has as its very core the objective of solving the root problems of the Middle East, will provide a forum at which both sides may reach a common understanding for their own national interests unencumbered by previous biases and influences. I do believe that the very international nature of this corporation can and should be used as a vehicle to accomplish what narrower national interests today prevent.

Creation of such a entity would provide a *non-national third party* with which both the Arabs and Israelis could talk.

As I have pointed out, the proposal that we have made far surpasses any partisan consideration. In espousing the idea, it is not my intention to try to preempt in any way executive responsibility nor is it my intention to try to involve myself in the negotiation of our foreign policy. For these reasons. I have chosen to try to crystalize broad support for the idea by introducing a resolution which would make it the sense of the Senate to endorse the idea and commend it to the Administration. I am very pleased that 52 of my colleagues from both sides of the aisle have chosen to cosponsor the Resolution.

Statement by General Eisenhower

Senator BAKER. I also have a statement from General Eisenhower, which is brief, and, with the committee's permission, I would like to read that into the record.

Senator SPARKMAN. Very well.

Senator BAKER. Mr. Chairman and members of the committee, reading from the statement of Gen. Dwight D. Eisenhower dated October 19, 1967, I quote:

I understand that the purport of the resolution offered by Senator Baker is an endorsement of a proposal to provide fresh water, electrical energy, and a longterm solution for the refugee problem in the Middle East. I am glad to be recorded as whole-heartedly approving the proposal and hopeful that the Resolution will be approved.

My acquaintance with the Middle East over some years impressed me with the critical importance of water in that area because of its very limited availability in a number of the countries there situated. As far back as we have written records, water, whether in wells, springs or small streams, has been the cause of contention, dispute and warfare, originally between families and tribes, and now between nations.

As a result of modern communications and modern weapons; warfare in any part of the world, however remote it may seem, is a potential risk to all the world. Therefore, self-interest as well as our common humanity warrant our concern with the unrest in the Middle East and its causes.

To the extent that disputes over an inadequate water supply have been responsible for, or have aggravated, animosities in that area, there has been until recently little that we could do to alleviate the condition. The Jordan River plan sponsored by our government in 1953–1954 was not accepted by the parties at interest. A large-scale program of well drilling suggests itself, but experience has proved that this is self-defeating for the rising salt water table, connected no doubt in some way with the Mediterranean, eventually contaminates well water when significantly large amounts are drawn from the water table.

As long ago as 1954, I asked the Chairman of the Atomic Energy Commission to investigate whether or not atomic energy would be applicable to large-volume desalting of sea water on an economic basis. At that time, however, there was no commercial plant producing power from a nuclear reactor, and the forecast was far from optimistic that the use of atomic energy as a source of heat either to simply evaporate water or to produce electricity would ever be competitive with the burning of coal, oil or gas. The first commercial nuclear power plant was begun in 1954, however, and though the cost of power upon its completion was not expected to be, nor was it, competitive, its success technologically encouraged others.

The succeeding years saw the erection of progressively larger installations and a concurrent reduction in the cost of kilowatts. I am told that within the past 18 months, more nuclear plants have been planned and contracted for than conventional plants.

The AEC scientists, notably at Oak Ridge, continued to work on the problem of desalting in cooperation with the other interested Federal agencies, and it is my understanding that they believe there are no unsolved scientific or technological problems in the way of

Construction of Nuclear Desalting Plants in the Middle East 67

achieving very large-scale production of fresh water from sea water and at a cost which, while of course not competitive with locations where adequate rainfall is provided by a bountiful Nature, is nevertheless of immense significance in situations where rainfall is sporadic, minimal, or practically non-existent.

Admiral Strauss, who served in my administration and the preceding administration on the Atomic Energy Commission, recalling our conversations on the subject, produced a proposal last spring designed to give effect to the great progress which has been made. The proposal appeared to me to be constructive and to breach a static situation.

Of course, a cardinal objective of the proposal is its insistence upon fruitful negotiations between the Middle Eastern beneficiaries of the plan and our presumption and hope that if they can be brought together to discuss and agree upon an allocation of the blessings which the proposal will provide them, they might then profitably discuss and negotiate the acute but less fundamental issues which presently divide them. Among these issues is, of course, the refugee problem to which we have been making large annual contributions both directly and indirectly. The provision of work and subsistence for these people may not be taken into account by those who will examine the economics of the proposal with a sharp pencil, but they are far from intangible aspects of its value.

I understand that testimony is to be given by Admiral Strauss and others on the details of the proposal. I would conclude by stating that I am enthusiastic about it and I believe that either in its present form or a modification thereof, it will bring a more abundant life to some millions of people and reduce the tensions from which wars are generated.

Statement of Hon. Stewart L. Udall, Secretary of the Interior; accompanied by Frank C. Di Luzio, Assistant Secretary of the Interior for Water Pollution Control

Secretary UDALL. Thank you very much, Mr. Chairman. I have at my right Assistant Secretary Frank Di Luzio, who was formerly head of the Saline Water Office, and now supervises desalting and water pollution control activities in the Department of the Interior.

I have prepared statement which I should like to go into the record. I shall read parts of it.

Senator SPARKMAN. You handle it as you see fit. The entire statement will be printed in the record.

Secretary UDALL. I should like to add some other matters that do not appear in my statement.

Senator SPARKMAN. Very well.

Bipartisan spirit of proposal

Secretary UDALL. Mr. Chairman, I should like to begin in the same tenor that Senator Baker did, and to add my own comments with regard to the very fine bipartisan spirit that has developed here this morning. I would like to say to the committee that I talked to President Johnson at length about this particular subject; I believe it was five weeks ago. He directed me at that time to communicate with President Eisenhower to inform him that he considered the plan, with which President Eisenhower had identified, and which is the subject of this hearing, a welcomed initiative, that he considered this a constructive idea, and that he hoped the two political parties could work closely on this without any hint of partisanship. I did so communicate this to President Eisenhower.

I can't think of two programs that have about the same lifespan where there has been less partisanship than saline water and the atomic energy programs generally. The Eisenhower administration initiated the atoms for peace, the saline water program was actually initiated in the last two years of the Truman administration. Both parties and four admistrations have given strong support to these programs.

We never made politics with them, and I think certainly this is the type of area, sensitive as it is in terms of international politics, where we simply need the best and most creative thinking that we can get. For this reason, I think this hearing is very timely.

I am delighted to see a Republican initiative on this, and I am delighted to see the keen interest of the Democrats. I can assure you that the Administration has a very keen and lively interest in this subject, as we shall demonstrate here this morning.

Political settlement a predicate to action

I want to put one caveat on my comments, Mr. Chairman, so that nothing I say this morning is misunderstood because I am here essentially as a resource expert. I am not a diplomat.

I think I know a little bit about politics but I don't pose as a diplomat and it is very obvious that with a serious and divisive political atmosphere existing in the Middle East today a political settlement is a prerequisite to action in terms of the type of plans and programs referenced in Senate Resolution tion 155.

I don't want any of my remarks to be interpreted as political meddling in a very sensitive area, but one does have to acknowledge certain things and I shall try to do that with delicacy. We are tremendously encouraged when we look at the resources of this region and at some of their potentials.

It so happens that only last February, I made a visit to this region. Secretary Di Luzio was with me. I did not, and may be in light of the present it is

Construction of Nuclear Desalting Plants in the Middle East 69

just as well, get to Israel or to the UAR. I was in Lebanon, Jordan, Saudi Arabia, Kuwait, and Iran. I think I got a pretty good picture of those areas. I helicoptered over most of the Jordan River Valley.

Involvement in resource problems of area

My Department has a long history of involvement in the resource problems of this region. The Bureau of Reclamation has done studies and work in the Jordan Valley. As has already been commented on, we have been working with Israel on this desalination plant that has been proposed for the last three or four years. We are just about ready to begin building a plant in Saudi Arabia, a water desalination and power plant.

Prior to the hostilities in June, I had an 11-man National Park Service team in Jordan assisting the Jordanians in setting up a national park system in that country.

Developing the arid lands

Secretary UDALL. Some of you in the Congress have seen the water desalination potential and this is the reason the desalination program has had strong support over the years. Where it will have the greatest impact in the world, in my judgment, is not in our own country, but in the arid lands of the world near the oceans, the rim of the Mediterranean, the Middle East region, where there is plenty of sunlight, where there is good soil, where it is near the ocean, and where previously there has been little rainfall. I think we realize that this is where the water desalination program in the next 50 years will have its greatest impact; there is no question about it.

This is going to be a very revolutionary development because water is the key. I need not remind members of this committee that the most productive agricultural areas by far in the United States today are the desert areas, where there is abundant sunlight. If you have water in California, in my own State of Arizona, and in the Southwest region, you can produce by intensive agriculture because you farm 12 months of the year. With your long growing seasons, with the type of agriculture that you can develop and with the many other favorable factors, this region of the world could literally bloom when one looks ahead during the next 100 years.

So this is something that to us working in the field of water desalination is very exciting indeed. When I saw the people of Saudi Arabia using merely underground waters that were rather brackish, and yet they were pushing back the desert and reclaiming it, this to me was the real wave of the future.

We certainly welcome the initiative expressed in Senate Resolution 155. It is a constructive approach to using water for solving critical international problems. Having personally visited the area earlier this year and having focused on the many problems presented, I am convinced that appropriate actions to provide water will raise the level of living and contribute to peace in that troubled area. I have also seen the great total resource potential that exists in the Middle East. The problem, then, is one of selecting and implementing the best approaches to these vast opportunities.

U.S. Government Cooperation with Middle Eastern States

First, let me say that the U.S. Government has cooperated closely with Middle Eastern States in applying developing technology to the area's water problems. On February 6, 1964, President Johnson noted that the United States "had begun discussions" with Israel on desalting. During Prime Minister Eshkol's visit to this country in June 1964, the two leaders reached agreement "to undertake joint studies on the problems of desalting". Subsequently a study of the feasibility of constructing a large-scale, nuclear-fueled, dual-purpose electric power and desalting plant in Israel was jointly financed by the United States and Israel. The study, completed in February 1966, the one that Senator Gore referred to, concluded that it was technically feasible to construct a plant which would produce 100 million gallons a day of desalted water and 200 megawatts of salable electricity. Further studies of the economic feasibility of constructing such a plant were carried out under the leadership of then Ambassador at Large Ellsworth Bunker.

In his speech at the climax of the Middle East crisis last June, President Johnson said:

> If the nations of the Middle East will turn toward the works of peace, they can count with confidence upon the friendship and the help of all the people of the United States of America... We here will do our share... and do more... to see that the peaceful promise of nuclear energy is applied to the critical problem of desalting water and helping to make the deserts bloom.

The United States—Office of Saline Water, Department of the Interior—has also cooperated with Saudi Arabia on the design of a fossil-fueled, dual-purpose desalting and powerplant for Jidda along the Red Sea. I was there to break ground on that last February.

Equipment is now being ordered for a plant which will cost $18 million and produce five million gallons per day of desalted water and 50 megawatts of electricity. Work is proceeding on schedule.

Senator COOPER. Mr. Chairman, may I ask a question?

Senator SPARKMAN. Senator Cooper.

Economic feasibility of proposed Israel plant

Senator COOPER. In referring to the study made for the plant for Israel, you say that it is technically feasible. Has it been economically feasible?

Secretary UDALL. Well, the big question, I would say, Senator, that is still on the table with regard to the Israel plant is the question of the economics of it and how you put it together. As I will show you in a moment, with the big southern California plant that we finally have put together, economics is the key. After all, water is something that man uses, it is a utility; man must pay for it. How much you charge for water, what your interest costs are, how it is financed, these become the key questions. This is the unresolved issue with regard to the Israel plant today. You are quite right.

Senator COOPER. Has any determination been made on that point?

Secretary UDALL. No. These are still hanging fire.

Economic feasibility of Jidda plant

Senator COOPER. My second question relates to the plant at Jidda. Was that plant found economically feasible or will the cost be assumed by the Government?

Secretary UDALL. Senator, this is a very interesting project.

Now, with Saudi Arabia, of course, the one thing they are so wealthy in is oil and gas. So when we got to the question of fuel, and fuel costs are one of your big elements, you see, in a water desalting plant, it was much cheaper for them to provide fuel oil, which they have in abundance. Therefore, we did not even look at nuclear energy because they have a cheaper fuel, and this helps keep the costs down.

It is also a plant that is small in size, and you go to nuclear power when you go to large sizes.

But this is a plant that the Saudis are themselves paying for. They have budgeted the money for it, although we are acting as a contracting agent, we are making a small investment in terms of a research investment. They asked us to come in because of our competence, and we are working very closely with them on this.

A market for electricity questioned

Senator GORE. Will the Senator yield? The same basic question that relates to water, the distribution of it, the use of it, the cost of it, and who is going to buy it and pay for it, relates equally to electricity.

Secretary UDALL. Yes.

Senator GORE. Where is to be the vast market for electricity of a huge dual-purpose plant? Like you I visited all these countries, and recently visited refugee camps on both sides of the controversy there.

The type of work for which the refugees are now prepared hardly relates to the use of atomic energy. It is more nearly agricultural type but water is necessary for that. But what do you do with the vast amounts of electricity? Who pays for that?

Secretary UDALL. Well, whether to couple together one plant to produce water and electricity is one of the basic decisions, Senator, that we made. I think the first year I was Secretary, we brought Jim Ramey of the Atomic Energy Commission, Dr. Weinberg of Oak Ridge, and others in on it, and we could see there would be tremendous economies if you did two things: One, you coupled one plant together that produced electric power, and secondly, you would bleed off the low-pressure steam from that and run it through a distillation plant to desalt water. We also determined if you went to large sizes you would get economies of scale, and these two things would drive the cost down because the whole exercise has been one of economics—how can we get the costs down.

We have always been able to produce desalted water, but can we produce cheap desalted water? That has been the task. You are quite right.

Regional cooperation is goal and prerequisite

Senator GORE. Basic to this whole proposal which Senator Baker and President Eisenhower both recognize in their statements is regional cooperation, regional use, and a prerequisite to that is some formula for the people of the area to live and work together. This is the goal as well as the prerequisite.

Secretary UDALL. Senator, I am a great believer in working together, and that is one reason why I was interested when I was in Iran, in finding that our diplomats and our people were not particularly disturbed with the economic arrangement the Iranians made with the Soviet Union to sell waste flare gas to the Soviet Union. It is my belief that one of the greatest forces for peace in the world today is interdependence and interconnection with regard to resources. The more pipelines, the more electric powerlines, the more water canals that connect countries, the safer this world will be, and this is the reason that—you are quite right in this region if you are going to generate enormous quantities of electricity, you must have an electric power grid. It will have to cross international boundaries. This will mean cheap power for everybody.

The benefit of interconnection of water power from large projects is that you get the end product cheaply, and if we can just get this one big thought across, "If you work together, if you will go to projects of large size you can have cheap power, you can have cheap water, and you can have all the economic benefits that derive from them".

Senator GORE. Thank you, Mr. Chairman.

Senator SPARKMAN. All right, Mr. Secretary, you may continue with your statement.

U.S. negotiations with Kuwait and Iran

Secretary UDALL. During my trip to the Near East last February, I expressed the interest of the United States in cooperating with Kuwait in a program of research in certain aspects of desalting and the training of operational employees envisioned by the Kuwait Government.

Just last summer we negotiated an understanding with the Kingdom of Iran to undertake joint studies of the water resources and requirements of Iran and to make recommendations for increasing water supply, through desalting techniques and other methods, to the end that the demand for agricultural, industrial, and domestic water uses can be met.

Opportunities for resource development projects in the Middle East

From my own experience in the area and my Department's work there, I would like to discuss briefly with you further the vast opportunities for developing water and other natural resources of the Middle East. These opportunities point up resource development projects for (1) creating jobs and landownership opportunity for refugees, (2) increasing agricultural productivity, (3) improving Israel-Arab relations, and (4) advancing the role of the United States as a peacemaker in the Middle East. These are the admirable goals of Senate Resolution 155 that can be attained through imaginative diplomacy.

I shall submit for the record data which have been formulated by the

Geological Survey, Bureau of Mines, Bureau of Reclamation, Office of Saline Water, and the National Park Service, all of my Department, with regard to different areas of activity in resource development work that we believe to be fruitful (see appendix). The following are some of the highlights:

1. Development and improvement of agricultural production is the first order of priority now, for more food is needed by local populations (for example, foodstuffs accounted for more than 30 percent of Jordan's imports in 1965).

2. Improved agricultural production requires:
   a. increased utilization of water, and
   b. better land use.

3. An integrated power grid is a must for the Jordan Valley and other Middle Eastern countries. Possible new power sources include:
   a. Hydropower, mainly local sources from pumped storage projects and from the Mediterranean-Dead Sea "siphon".
   b. Nuclear power, including combining power and desalting plants.
   c. Oil and gas.
   d. Oil shale, Jordan and Israel have rich bituminous limestone and shale deposits.

4. New or expanded mineral industries can be developed, including chemicals, fertilzers, ceramics, construction materials, abrasives, and metals.

5. Considerable economic potential exists in processing agricultural products, fruits, oils, leather, and canned foodstuffs.

6. Economies can be bolstered by revenues from tourists interested in properly preserved holy places, antiquities, and the scenic values. Just prior to the outbreak of hostilities, the National Park Service of my Department had an 11-man team in Jordan to help develop this vast potential. Other Jordan River Valley countries have similar underdeveloped resource potentials.

Solution of Jordan River water problem

The solution of the Jordan River water problem and the displaced persons crisis lends itself to a comprehensive regional solution when we look at this as a resource problem. The first step would be the development of the Jordan River to meet its maximum capability. Attention should then be directed to supplementing the naturally available Jordan River water supply in other ways. We can talk about augmentation today, not just using what you have, what nature has put in the river, but augmenting it.

Based on present potentials, the required water needed in addition to the natural runoff of the Jordan River has been estimated to be in the range of 500,000 acre-feet annually. The combination of maximum planned development by Israel and Jordan exceeds the available supply by 300,000 to 350,000 acre-feet annually, and Syrian plans contemplate additional annual use of about 100,000 acre-feet, so that you can see the overdemand already that exists in terms of the future.

Weather modification procedures

There is every reason to believe that some of this additional water need can be provided by weather modification procedures. My scientists tell me that the headwaters region of the Jordan River offers a topographic situation favorable for the formation of clouds suitable for cloud seeding. During the winter months, potentially unstable air frequently flows across the region. Temperatures aloft in this moist air are characteristically cool enough to justify seeding.

Within a year after I became Secretary of the Interior, the Bureau of Reclamation intitiated its atmospheric water resources research program with an appropriation of only $100,000. After only six years, the appropriation for 1967 is $3.75 million. Future projections for the nationwide engineering and scientific research program to "tap the rivers in the sky" as we call it, contemplate expenditures in excess of $50 million a year over a 10- to 20-year period. Our scientists tell us that we will be ready before 1975 for full-scale field application on a river basin like the Colorado in the West, a river that is short, just as the Jordan is short. We can increase the average annual runoff by 10 to 20 percent—these are conservative figures—and if this can be done in the western part of the United States or elsewhere in this country it can also be done in the Jordan Valley.

The objective of this new program is to develop operational systems to augment precipitation, when atmospheric conditions are right, and to store and regulate the additional supply for use during dry periods.

Current technology indicates that successful seeding of the Jordan River headwaters could be expected to generate about 100,000 acre-feet of new water annually at a very modest cost. In fact, we figure the cost at 50 cents an acre-foot, which is very cheap and far below the cost, for example, of desalted water. The actual seeding operation, however, would probably have to be conducted with the knowledge and consent of the nations of Lebanon, Syria, Israel, and Jordan, since it might conceivably affect all four.

Ground water and water reuse

Ground water potentials are nearly fully developed in Israel, according to our analysis. Ground water has helped build an agricultural base prior to construction of the national water network. Ground water can provide an interim solution to water problems pending long-range development of other sources because it can be obtained without international agreements and it is a low-cost, short-term supply. A considerable ground water potential exists in eastern Jordan.

Although the people of the region have made tremendous strides in efficient use of water, further development and training in water reuse and conservation would help solve the problem.

Status of plans concerning desalting

I have already mentioned some of the work we are doing with Middle Eastern countries in desalting. I should now like to provide this committee with as concise a picture as possible on the status and plans concerning desalting, especially in view of the reference in Senate Resolution 155 to construction of desalting plants for providing large quantities of fresh water to both Arab and Israel territories.

With wisdom, vision, and foresight, the Congress in 1952 authorized the Secretary of the Interior to conduct a research and development program to develop low-cost means of producing from sea water, or other saline waters, water of a quality suitable for agricultural, industrial, municipal, and other beneficial consumptive uses. This was a dramatic new concept for until that time desalted sea water had been used only as a method of obtaining fresh water for ships at sea and in a few isolated arid ares of the world.

The appropriations for the Office of Saline Water program in fiscal years 1964–67 totaled over $80 million. This is more than double the total funds which went into this program in the entire 12 preceding years. The program of this Office is summarized in material I shall furnish the committee.

As early as May 1957, the Office of Saline Water awarded a contract to the Fluor Corp. of Los Angeles, Calif., for a study of the applicability of combining nuclear reactors with saline water distillation processes.

Bolsa Island project

Much of the new technology developed over the brief span of time that the saline water conversion program has been in operation will culminate in the

Bolsa Island nuclear power and sea water desalting project. The project gets its name "Bolsa Island" because it will be constructed on a 43-acre manmade island, that does not exist today, about one-half mile offshore from Bolsa Chica State Beach in Huntington Beach, Calif.

Mr. Chairman, I should like to say that, this is the model we have here. It is far and away, of course, the most exciting, largest desalting plant in the world. The capacity of this plant is many times larger than the total desalted water that is being produced in the world today.

This is the result of three years' work. It took us three years in terms of the negotiations that put all the pieces together. It includes two main features. Over here you see two nuclear reactors, very large, modern reactors, each with a capacity of 715 megawatts—Commissioner Ramey will explain how we get 100 more each—so there is a total of about 800 megawatts net to constitute one of the largest, most modern electric power generating plants in the United States or in the world. This plant will be built by the electric power utilities in southern California, both public and private. It will be tied into the grid network of the city of Los Angeles and Southern California Edison Co., San Diego, benefiting the entire region. The Atomic Energy Commission will make a modest contribution to help make that plant possible.

Water desalting project

We then move over to the water desalting part of the project. It will be built and financed by the Metropolitan Water District of Southern California, which is the largest water wholesaling entity in the world. The Federal Government only participates by making contributions, by helping with the engineering. The water will be put into the southern California system that runs all the way from the Colorado River to San Diego.

These three banks you see will each produce 50 million gallons a day. To give you some idea of sizes, I have said the Saudi Arabian plant we are building will provide five million gallons per day. There are plants in Kuwait today that are either in existence or are being built to produce roughly five million gallons per day. Each of these three banks produces 50 million, a total of 150 million gallons per day.

This has been authorized and the contracts are before the Congress at this moment; we expect to sign them next month. The Federal Government will contribute roughly half the cost of the first unit—the first 50-million-gallon-a-day unit which you see here.

We are essentially buying technology. We are putting in $61 million, and we are, in effect, buying for the United States the technology, the knowledge, that we will get by having this built and having it operate. Basically the Metropolitan Water District will build the other units without much Federal assistance.

This, of course, will yield some of the cheapest power that is produced anywhere in the world. The million-gallon-a-day demonstration plants we have in this country today are producing water for roughly a dollar to $1.25 per thousand gallons. This plant, with the economies of its scale, with the savings because we take the low-pressure steam from the nuclear reactors and run it through the desalter will produce water, we calculate, at 22 cents per thousand gallons. In other words, we have broken the cost from, let us say, $1.25 to 22 cents.

This 22-cent cost is very close to the cost for which the Metropolitan Water District will get water from the California State waterplant from northern California, bringing it down through a series of canals to a pumping station.

So we see we have had the economics put together properly. We had to have a project where the financing was there, where we had a market for the electricity, where we had a market for water, and where the users of electric power and of the water will essentially pay for this plant and amortize it in the usual public utility fashion.

Costs estimated in Kaiser study for Israel Government

Senator GORE. Mr. Secretary, I would like to cite as of comparable pertinence that you estimate the cost of water here as possibly as low as 22 cents for 1000 gallons. This feasibility study to which I earlier referred and to which you also referred, by Kaiser engineers, estimates the cost of water in this Middle Eastern study at 43.4 cents per 1000 gallons at a 7-percent interest rate charge, and as high as 67 cents per 1000 gallons with a 10-percent interest rate charge, which is rather a low interest rate charge in the Middle East.

Senator LAUSCHE. Senator Gore, to which plant or project does that specifically refer?

Senator GORE. This is an engineering feasibility and economic study for the dual-purpose electric power and desalination plant that is under study for Israel.

Secretary UDALL. It is the Kaiser study for the Israel Government.

This is a very key question. I would like Secretary Di Luzio to tell you the difference in the costs, Senator.

Mr. DI LUZIO. I think the difference in the costs is due to three factors. One, of course, is the size of the reactor for Israel.

The reactor size of the Israel plant is 300 megawatts; prior to being raised it was 200 megawatts; and these, of course, are much larger, with the result that lower energy costs are coming out of these reactors.

The second factor involved here is the cost of money. The 7-percent factor is based upon the availability to Israel of certain kinds of money whereas the California plant has two kinds of money in it: that put in by the city of Los Angeles, a public utility, which operates on the sale of bonds, very low interest bonds, and the private investment from the two private power companies whose cost of money is much higher than the city's. The city of Los Angeles owns one of the two reactors and is contributing steam from that reactor to the plant.

The third factor is the fact that the 22-cent cost we talk about, 22.9 actually, in the report itself, is this total cost with nothing written off, including Government contribution. That is the total cost of the plant itself, and it is not the cost of this water to the Diemer filtration plant, which is the beginning of the distribution system.

Senator GORE. What would that cost be?

Mr. DI LUZIO. That would be about 4 cents per thousand more, just to convey the water over, so there are both financing differences and the scaling effects of the two plants.

Developing a market for electric and water power

Senator GORE. Plus the factor that in the case of the plants at Los Angeles you have a readymade demand for the power, whereas in the Middle East, because of the hostilities there and inability under present conditions to build transcountry networks, you do not have a market and would not have a market until and unless we could obtain, or they could develop, some peaceful formula for coexistence and cooperation; is that correct?

Mr. DI LUZIO. Well, it is correct. But I think, Senator, if I might suggest something, that it is going to take some dramatic act such as this to convince them that there is economic gain for all if they join hands because the Jordan cannot be developed by any single country.

Senator GORE. That is why I said a few moments ago that inherent in this resolution which my distinguished colleague has put forward, and which General Eisenhower and Admiral Strauss have recommended, is that this would be both a goal and a prerequisite, and an incentive to attain these kinds of conditions, but without which the plant would not be possible. You concur in that?

Mr. DI LUZIO. I agree with that, and I think how one phases the phasing of the program is the key. How do you phase it? Obviously you do not go for all three plants or any two plants by themselves initially. I think what is critical is the timing, and to do first things first; start discussing the economic benefits of the program if they will by joining hands, and start immediately doing certain things necessary to settling down the refugee problem which, I think, is the heart of the whole problem. The political problems which you mentioned, Senator, while true, are a little bit offset by the fact that the refugees are floating between both poles now. They are in the center of this conflict, and both sides know they must do something about the refugee problem if they are ever going to settle the Middle East situation at all.

The west bank has tremendous potential as an agricultural and commercial area and can be developed without really much regard, or perhaps very little regard, as to whether Israel has jurisdiction or Jordan has jurisdiction. It is facing up to the refugee problem which, in my opinion is the first step.

Economical price for water in mideast area

Senator LAUSCHE. It is mentioned in the study on the project of which Senator Gore spoke about, that at an interest rate of 5 percent, the water cost would be 28.6 cents. If the interest rate is 7 percent it would be 43 cents; if the interest rate is 10 percent, it would be 67 cents.

What have you concluded to be the price that can be paid for water economically in that area? Will this be self-sustaining on the basis of the costs that the water will have to bring?

Mr. DI LUZIO. I cannot answer the second part of your question, but I can answer the first part. The first part is that the Israelis feel they can take water costs up to something around 30 or 33 cents and still make a positive contribution to their agricultural products.

Senator LAUSCHE. But if it goes beyond 33 cents it has to be subsidized?

Mr. DI LUZIO. It has to be subsidized.

Senator LAUSCHE. Or money borrowed at a reduced rate of interest that will bring the costs down below 33 cents per 1000 gallons.

Mr. DI LUZIO. Or you must gain something else by the utilization of this water beyond pure economics.

Locations of proposed plants

Senator LAUSCHE. For what three cities or countries are these desalination plants contemplated? One in Israel, one in the United Arab Republic. Where is the third?

Mr. DI LUZIO. Well, Saudi Arabia is the second one. There are two in Saudi Arabia, and others we have been discussing with Morocco, Algiers, Tunisia. I think the plants proposed by Admiral Strauss are two in Israel and one in Jordan.

Senator LAUSCHE. Is this one in Saudi Arabia at all involved? The Department of Interior was in on that.

Mr. DI LUZIO. No, those are two small plants.

Senator LAUSCHE. Is the one in the United Arab Republic at all involved?

Mr. DI LUZIO. No

Senator LAUSCHE. These two are.

Secretary UDALL. These two are not involved in the plan with which President Eisenhower is identified. The Saudi Arabia plant is being built, Senator. We are going forward with that.

Senator LAUSCHE. The one Senator Gore has been referring to down at the Gulf of Aqaba is involved.

Mr. DI LUZIO. No. The two plants in the Gulf of Aqaba which we referred to are existing plants; the first one used the Zarchin process, which was a freezing process, built by Fairbanks Morse. The second one was a Baldwin-Lima-Hamilton evaporator which the Israelis themselves paid for. There are already two plants now in the Gulf of Aqaba, one a million gallons, and the other a million and a half gallons.

Overhead irrigation of nonintensive culture crops

Senator GORE. Mr. Secretary, this may be irrelevant, but I doubt that it is. When I was in Israel a few weeks ago I noticed a considerable amount of overhead irrigation of nonintensive culture crops such as cotton.

Now, my own experience has been that this is uneconomic even in our own country unless you can irrigate by the natural flow of water, in other words, surface irrigation; that the vast amounts of electricity and piping and labor investment involved in overhead irrigation is so expensive that it can only be justified with an intensive culture, and that does not include cotton and it does not include corn, hay, and pasture.

Would you generally agree with that?

Secretary UDALL. I think you are certainly right in terms of your analysis, Senator. I do not know, I am not familiar with, the particular situation you describe.

Senator GORE. I certainly would not want to see us subsidizing a type of irrigation which we find uneconomic even at home with our own types of electricity rates.

Secretary UDALL. I want to make one point because I think we are very close to what the big decision will be that this committee and the Congress will face if these people can solve their political problems enough so that you can get this kind of planning and work started.

I think you are ultimately going to get down to the question that you put your finger on here of whether the international community is willing to subsidize the difference to make such a project possible.

This project would not have gone forward—the sticking point on this project, and, we fought over it for nearly a year with the southern California people, was this $72 million—$61 million to build plus $11 million operating and maintenance costs—that the Federal Government had to put in, otherwise the Metropolitan Water District and the cost of the water would be too high. They said, "We can get water from northern California."

We had to make it competitive with water economics of southern California, and personally I think the best money that the world could spend for peace today is to put together whatever element of subsidy is necessary to make the project possible in the Middle East. This is ultimately what the world community will have to do because you are talking about water and electricity. People use them, they have to pay for them, so you have that fact of economics which is in your favor. But if the cost is too high you are going to have to have an element of subsidy. But we, every week in this city and in the world, vote more money for military expenditures, and we never question it at all. Maybe a small investment here would mean that the whole world does not have to make military expenditures some time in the future, and that is the reason I think this is probably, whatever it is, the best money we will ever spend in terms of buying peace.

Senator GORE. No more questions.

Secretary UDALL. Senator, I have been interrupted, and have not finished my statement, and I will rush right ahead, if I may.

Senator SPARKMAN. Go right ahead.

Plans for desalting plant

Secretary UDALL. The Bolsa Island desalting plant will produce 150 million gallons of water per day, sufficient fresh water to supply the needs of a city the size of Boston or San Francisco; and the turbine generators will produce 1800 megawatts of power which would serve the electrical requirements of approximately 2 million people. As fantastic as this may seem, we already are considering this plant only to be a prototype of plants to follow. In 1965, the United States, Mexico, and the International Atomic Energy Agency entered into an agreement for a preliminary assessment of the technical and economic practicability of dual-purpose nuclear power and water desalting plants to produce fresh water and electricity for the States of California and Arizona in the United States and the States of Baja California, and Sonora in Mexico. A final meeting of the team that has been formed to conduct this joint study is underway this week in San Diego, Calif. We are in the early stages. The size of the basic desalting plant under construction is one billion gallons per day, in other words, nearly 10 times as large as this one.

I can think of no other commodity that would do more to quench the fires of war and tension that exist in the Middle East than an abundant supply of fresh water. The technology to supply that water is within our grasp today.

Thrust of Department's desalting program

Since its inception, the entire thrust of the Department's desalting program has been develop lower and lower cost methods of producing fresh water from saline sources. While remarkable progress has been achieved some still contend that this water is still not yet cheap enough to use for irrigation. But the use of desalted water for irrigation in the Middle East cannot actually be compared to existing irrigation projects anywhere else in the world. We have reached a point in time when it has become necessary to consider the value of water rather than its cost or price, and the three terms are not synonymous. I have read that the cost of the recent 6-day war in the Middle East exceeded $2 billion. If a few hundred million dollars investment in a desalting plant could in any way contribute to the prevention of a future conflict, it may indeed be the wisest investment of our time.

As President Johnson has said: "We are not using all the imagination and all the enterprise that our problem requires."

We do not advocate that a plant be built to raise agricultural products which are readily available in other areas; but provided with the lifegiving water, I am convinced that the fertile lands of the Middle East could produce high-value cash crops to meet the rising demand fo food and fiber throughout the world. With the construction of desalting plants in the Middle East, we will need to develop better water conservation practices, select productive crops, learn new irrigation techniques, organize new and effective methods of water management, find better ways to fertilize the soil, and many other methods of extending the use of water to its full potential.

We must do all in our power to apply 20th century technology to projects for peace. Nothing I have suggested is beyond our technical capability; all that is needed is the willingness of nations to join hands to advance such mutual projects to their completion.

Planning could begin today. For the initial phases of construction, the same design that is being developed for the Bolsa Island project could be utilized.

So when should construction begin? Would it be wise to wait until all of this technology is in hand? Or should work begin now that the need is so clearly demonstrated?

What can be done technically with desalting

Technology available today in desalting makes it possible to consider a wide variety of plant sizes and processes to meet the supplemental water requirements existing in the Middle East, including agricultural, municipal, and industrial needs. We deliberately tried to develop a very versatile program of large plants, middle-size plants, and small plants to meet all needs. Desalting plants can be built in capacities to meet conceivable short-term and long-term water requirements of the population, providing for the needs of the thousands of refugees and displaced persons. Dependent upon the requirements, it is possible to build single-purpose waterplants only or dual-purpose water and electric powerplants utilizing fossil fuels as well as nuclear power. Someone has even suggested, and these are other alternatives you can look at other than nuclear power, that you could pipe in the waste flare gas from the Arabian Gulf area. Such enormous waste of energy is one of the things that strikes a conservationist in a very negative way. You could probably buy it at very cheap cost, pipe it in, use it as energy, and at a later stage go to nuclear energy and use the gas for petrochemical industry plants. There are any number of possibilities that this rich region could look at, if you take a total resource view.

Construction of Nuclear Desalting Plants in the Middle East   85

Friday, October 20, 1967

UNITED STATES SENATE,

COMMITTEE ON FOREIGN RELATIONS,

Washington, D.C.

The committee met, pursuant to recess, at 10:10 a.m., in room 4221, New Senate Office Building, Senator John Sparkman, presiding.

Present: Senators Sparkman, Gore, Symington, Clark, Pell, Hickenlooper, Aiken, Case, and Cooper.

Senator SPARKMAN. Let the committee come to order, please.

The purpose of the meeting this morning is to continue hearings that were started yesterday on Senate Resolution 155. Our first witness this morning is a gentleman who certainly can be helpful to us in considering this question, former Chairman of the Atomic Energy Commission, Adm. Lewis L. Strauss. We are glad to have you. If you will proceed in your own way. You understand, of course, that your statement will be printed in full in the record. You may read it, discuss it, summarize it, handle it as you wish.

Mr. STRAUSS. Thank you, Mr. Chairman.

Senator SPARKMAN. We are glad to have you.

Statement of Lewis L. Strauss, Former Chairman of the Atomic Energy Commission

Mr. STRAUSS. It is a great pleasure to be here.

Mr. Chairman and gentlemen, the testimony here yesterday by Secretary Udall and Commissioner Ramey, and Secretary Di Luzio was most encouraging to me. It certainly demonstrated that the subject before you is in no sense a partisan issue, and no defense against that charge is needed. If, as I profoundly believe, it is an idea whose time has come, what is needed now, I submit, is action.

I brought with me a statement briefly explaining the origins of the plan. It was drafted months ago before the war in the Middle East, and as you will see it is but a poor composition, the bare bones of the project. Its purpose, to bring the belligerents together now and eventually to assure peace in the area and a better life for those who live there.

It also aims to relieve us of military aid and refugee relief expenditures in the Middle East which, under existing conditions, seem to me might go on endlessly.

At the hearing here with me today is the Director of the Oak Ridge National Laboratory of the Atomic Energy Commission. As some members of your committee known, Dr. Weinberg's service there dates from the days of the Manhattan project in World War II. He has served under every Chairman of the Atomic Energy Commission, and aside from his invaluable work in many areas there, much of it still unpublished and unpublishable, he and his colleague, Dr. Philip Hammond, and their colleagues are the most original innovators and the best informed sources on the arithmetic—the scientific and technological aspects of desalting water with the aid of nuclear power.

Mr. Chairman, I will hasten through my statement in order that you may not be delayed in hearing from him.

May I read the statement, sir?

Senator SPARKMAN. Yes, you proceed as you see fit.

Inadequate supplies of fresh water

Mr. STRAUSS. Although a shortage of sweet water has begun to threaten some areas of our country, we have been so blessed with natural resources that an abundance of water for human consumption, agriculture, and industry has always been taken for granted, like the air we breathe. Pollution, it is true, has now begun to contaminate both our air and water, but we have a few years of grace to correct abuses which, unchecked, could make life on our portion of the planet intolerable.

There are parts of the world, however, where an adequate supply of fresh water has always been lacking, and this is glaringly true of a large area eastward of the Mediterranean between latitudes 30° and 34° north. From the most ancient times in those lands now divided between Israel and her Arab neighbors, water has been a rare and treasured resource. The earliest occupants of the country guarded their individual wells and springs with their lives, and battles were fought to possess or to retain them.

With the increase of populations in recent decades, contention for the available water resources has also increased. Scientific hydrology and modern well-drilling have tempered tensions, but only temporarily as the point of diminishing returns has been reached in the search for new sources of water. Thus, even though not the proximate cause of the recent hostilities, shortage of water is a fundamental basis of rivalry, hatred and war in the Middle East.

Construction of Nuclear Desalting Plants in the Middle East 87

Populations unable to grow enough to feed themselves are obviously unable to produce a surplus for export and thus create wealth upon which to base improved living standards and peace.

Diplomacy and military measures alike have proved ineffective to transform a situation which only lifegiving water in great quantities can alleviate. This was true in the days of the Bible, and there has been no change in the intervening millennia. "Any source of water", wrote Nelson Glueck, a distinguished archeologist familiar with every mile of the region, "however small and brackish, as long as it is drinkable, is accounted precious ... Even chance pools collected in hollows or wadi beds after the brief winter rains are not neglected. I have seen modern Bedouins pitch their tents by them until nothing was left but a rapidly drying patch of mud ... Men and tribes fight bitterly over wells and cisterns."

Jordan River project

The situation in 1967 is as it has always been—apart from rainfall so seasonal that for more than half of the year there is no rain, the water supply of the region comes from wells, a few insignificant streams, and from the Jordan. That river, by our American standards, as Senator Gore observed yesterday, is no more than a creek, its flow being about one percent of that of the Nile. A proposal for its equitable and wise distribution, worked out between 1953 and 1955 by the late Eric Johnston, was denied approval by the Arab countries on political grounds although the technical aspects of the plan had the endorsement of their engineers. Much of the Jordan still runs down into the Dead Sea, there to waste its once sweet waters in that poisonous brine.

On June 16, 1954, President Eisenhower, who had sent Eric Johnston on his mission, wrote me a note which read in part:

> I can think of no scientific succes of all time that would equal this [desalting of water] in its boon to mankind—provided the solution could do the job on a massive scale and cheaply.

The two criteria which the President established—"massive scale" and "cheaply"—were at that date not yet within the reach of technology. The simple desalting of sea water by distillation presented no problem in small batches. It was the regular procedure on many ships and shore installations where the amounts required were measured only in thousands of gallons per day and cost was not a severely limiting consideration. Water to drink can bear a relatively high cost in cents per gallon, whereas for industry and agriculture it would have to be produced in millions of gallons and at a cost of much less than 1 mill per gallon. Nuclear energy seemed to point a way to the realization of this goal but not at that time a clear way, for in 1954, the cost of thermal kilowatts derived from nuclear reactors far exceeded the cost of energy derived from the combustion of coal or oil. Furthermore, there were no commercial installations. In 1953 when I became Chairman of the Atomic Energy Commission the Joint Committee on Atomic Energy had just published a report on the subject in which they stated that there was, at the end of 1952, no commercial atomic powerplant either in existence or planned.

Cost of fresh water is diminishing

Soon, however, it became apparent that by scaling up the size of nuclear powerplants and by improving the design of the nuclear cores there would be a marked reduction in the cost of the output. And eventually, by 1961, using heat from a nuclear reactor to evaporate sea water, it could produce distilled sweet water at about $1 per thousand gallons. Barely five years later, by continued improvements, in the design of nuclear fuel elements and by enlargement of plants, the cost per thousand gallons of fresh water has been calculated at a little over 21 cents per 1000 gallons, a reduction of four-fifths. The dome which was here yesterday and which was an exhibit provided by the Atomic Energy Commission, is a model of the plant whose estimated cost of fresh water I have just cited. New reactor principles and the development of desalting procedures by methods more sophisticated than simple distillation will further lower the cost of fresh water.

On August 13, 1958, addressing the General Assembly of the United Nations, President Eisenhower dealt with the crisis in the Near East where, by the prompt dispatch of our armed forces to Lebanon, a catastrophe had been averted. Then, speaking of the great surge of Arab nationalism, he recommended a program for the development of the region, and high on the agenda of that program placed "action to solve one of the major challenges of the Near East, the great common shortage—water. New horizons", he said, "are opening for the desalting of water. The ancient problem of water is on the threshold of solution: Energy, determination and science will carry it over that threshold."

"A Proposal for Our Time"

With these two statements of 1954 and 1958 in mind, General Eisenhower had before him during the first week of this past June a memorandum which I had written entitled "A Proposal for Our Time" as follows:

> Attention to the debates in the United Nations since the end of May must convince the observer that an end to the trouble in the Near East is not in sight. The introduction of a new and dramatic element will be required to establish a climate in which peace can begin to be negotiated. The "lie direct" has been exchanged so often in the General Assembly that men can hardly be expected to reach agreement by rational discussion in the atmosphere which has been created.
>
> The two fundamental problems in the Near East are (a) water and (b) displaced populations. It is these issues, basically, which have exacerbated relationships in that area over the years, and it has been proved that they are not to be resolved by political or military measures. But by a simple, bold and imaginative step, it is in our power to solve both problems.
>
> Let a corporation be formed with a charter providing that the Government should subscribe to one-half the stock, the balance to be offered for public subscription in the security markets of the world. The amount thus to be initially raised, say $200,000,000, would be used to begin construction of the first of three large nuclear plants for the dual purpose of producing electrical energy and desalting sea water, with emphasis on the latter purpose.
>
> Two of the installations would be located at appropriate points on the Mediterranean coast of Israel and a smaller one at the northern end of the Gulf of Aqaba in either Jordan or Israel, as the most suitable terrain may dictate. Design and construction contracts would be let on bids in the several countries which have had experience in building large nuclear reactors, i.e., the United States, the United Kingdom, France, Canada, and, perhaps "under proper conditions" the Soviet Union.

Financing of proposed plants

The cost of the plants, beyond the sum raised by subscription to the common stock, would be financed in succeeding steps by an international marketing of convertible debentures. These would bear no interest for the first years (approximately the time required to construct the first plant), and could be income debentures depending upon market conditions at the time of issue.

Operation of the plants could be made the responsibility of the International Atomic Energy Agency, of which agency each of the major belligerents, fortunately, is a member, and that agency would also have jurisdiction and control of reprocessing of fuel elements to insure that all nuclear material was accounted for. The International Atomic Energy Agency would act as trustee for the owners of the stock and would dispose of the power and water, and service the securities with the proceeds.

During the Eisenhower Administration, the United States allocated for peaceful uses overseas considerably more than sufficient uranium to fuel these installations, and I have not heard that it has been withdrawn. The first plant would be designed to produce daily the equivalent of some 450,000,000 gallons of fresh water (incidentally, I am informed, more than the combined flow of the three main tributaries which make up the Jordan River north of the sea of Galilee). It would also produce an amount of power which, though well in excess of the present needs of the area, would attract industry, because of its low cost. Industries such as aluminium and heavy chemicals normally seek low-cost power and build to take advantage of it. It will also be profitable with cheap power to make the nitrogenous fertilizers needed in the Middle East and beyond. A large export trade would result from this item alone.

The power would be used to pump fresh water into the water-starved areas of Israel, Jordan and other Arab countries—perhaps even including part of Egypt east of the Nile Valley. (Presently, 20 percent of all the power produced in Israel must be used to pump its limited supply of water.)

The introduction of fresh water from all three plants into the arid and semi-arid areas would have the effect of opening to settlement many hundred square miles, part of which in ancient times supported now vanished settlements, of considerable size, and other areas which heretofore have never supported human life (other than on a nomadic basis). The blood-stained controversy over the division of the waters of the Jordan would thereupon become de minimus.

The work of building the great plants, laying the pipelines, constructing holding reservoirs (to allow for temporary shutdowns for maintenance), power lines, irrigation ditches, etc., will absorb the unskilled labor of scores of thousands of displaced persons and the skilled work of many hundreds. When the plants are in operation, the greater part of the labor force could be settled in newly irrigated areas under condition far superior to any life they or their fathers have ever experienced.

Solely as a measure of magnitude and without attempting invidious comparison, it might be noted that the completed project will represent substantially less than one-third of our annual expenditure on the space program. It should return its cost and then yield income in perpetuity, not only retiring the borrowings incurred but rewarding the governments and individuals with vision enough to subscribe to it initially.

Cooperation of Arab and Israeli Governments

Cooperation of the Arab and Israeli Governments will be necessary in order to agree upon a modus vivendi for allocating water and power, and it will be apparent that any government which simply declined to discuss it would have to answer to its citizens sooner or later. Water is an eloquent advocate for reason.

Where the President of the United States to electrify the world by such a proposal, as President Eisenhower did in his Atoms-for-Peace speech to the United Nations in 1953, it would be hailed by millions who now can see no way out of the morass in which the powers are presently floundering. The threat of triggering more widespread war is inherent in the situation as it now stands, while the proposal might well be the beginning of peace and a new life in these ancient lands.

The proposal, of course, does not settle boundary disputes, shipping routes and other acute but temporary issues now confronting the belligerents, yet settlement of these would be immensely accelerated by the pressure from all sides to get ahead with such a project where delay can only be counted in human lives and misery. It could be announced by the President that no affirmative steps would be taken until negotiations at least began.

In the atmosphere that would follow such a proposal, the leaders of the Near Eastern countries could be invited to come together on the basis of the proposal and for no other immediate purpose. Happily, they have a common and, thus far, unembarrassed forum in the International Atomic Energy Agency.

Private capital in both the United States and Great Britain would, I am sure certainly respond to the challenge of such an enterprise on the initiative of the Government.

Criticism of the proposal

Critics of the proposal will argue that the cost of desalted water, if it exceeds 15 cents per thousand gallons, will be too high for agriculture—though much farming can use water at substantially more than this figure. There was an identical cost objection by those who once counseled the Atomic Energy Commission that the development of electrical energy from nuclear fission would never be economic. Although that advice came from respected scientific quarters, it proved to be overpessimistic. It was found that the cost per kilowatt-hour fell very rapidly as the size of plants increased—from an estimated 18 mills per kilowatt-hour in 1949 to 2.6 mills, which the supposed, the estimated cost, of the plant we had described to us yesterday, and the cost is still going down.

Within the last year, nuclear power has become fully competitive with power conventionally produced, and I am advised that about half the new power plants contracted for in the U.S. in 1966 are nuclear. The same factors will lower the cost of fresh water.

That is the end of the memorandum.

General Eisenhower communicated it to President Johnson and the testimony we heard yesterday indicates that the proposal has had the careful consideration, and I believe favorable consideration, of the Administration.

The ideas expressed in my memorandum have undoubtedly occurred in one form or another to others concerned with the threat to world peace in the Near East, and if indeed there are many versions of the idea, it may be evidence of the truth that nothing can stop an idea whose time has come. It would be an inspiration for this generation and for those who come after us to see the vision of Isaiah realized:

The wilderness and the parched land shall be glad, and the desert shall rejoice and blossom as the rose.

Gentlemen, that is the end of my formal statement.

List of objections to plan

I have also set down a number of objections to the plan. These have come to me mostly by mail as a result of the many favorable editorials that the proposal has stimulated across the country. There is an answer to each objection and following Dr. Weinberg's testimony, if they have not been largely answered in your own minds, you may wish to have some of them developed. The list is as follows:

First. The United States has no real interest in the area.

Second. The proposal is too bold and too large.

Third. The proposal is too small, and will not help enough people.

Fourth. So large a plant has never been built before. It is too risky.

Fifth. To make that much fresh water there will be a great excess of power—uneconomic.

Sixth. You will not be able to borrow money cheaply enough.

Seventh. The refugees in question are by now pauperized and no good for work.

Eighth. The proposal will put weapons materials in the Middle East and proliferation will result.

Ninth. The International Atomic Energy Agency will be neither willing nor able to undertake the management of the project.

Tenth. No other countries or its nationals will help.

Eleventh. The Arabs and the Israelis won't be interested.

Twelfth. The cost of the water will be too high.

Thank you, Mr. Chairman.

Reference

1. Before the Committee on Foreign Relations, United States Senate, Ninetieth Congress, First Session, on Senate Resolution 155. A Resolution to Express the Sense of the Senate Concerning a Means Toward Achieving a Stable and Durable Peace in the Middle East. October 19, 20, and November 17, 1967. Government Printing Office, Washington, 1967.

   The reprinted portion is the whole of Senator Howard H. Baker, Jr.'s direct statement, including the statement from General Dwight D. Eisenhower which Senator Baker put in the record; selected parts of the statement by Secretary of the Interior Stewart L. Udall; and the direct statement by Lewis L. Strauss in its entirety.

VIII

NUCLEAR ENERGY CENTERS–AGRO-INDUSTRIAL AND INDUSTRIAL COMPLEXES[1]

Oak Ridge National Laboratory

Prefatory Note

Alvin M. Weinberg, Director

Oak Ridge National Laboratory

ORNL-DWG 67-11464A

CUMULATIVE NUMBER OF PROCESSES

- GASOLINE FROM COAL
- GENERAL-PURPOSE INDUSTRIAL HEAT
- PIPELINE GAS (H_2 + CO) FROM COAL
- ACETYLENE FROM CaC_2
- AMMONIA (AND HNO_3, NH_4NO_3, AND UREA)
- IRON BY ELECTROLYTIC H_2 REDUCTION
- ELECTRIC FURNACE PHOSPHORUS
- OXYGEN VIA AIR LIQUEFACTION
- CAUSTIC-CHLORINE
- ALUMINUM
- MAGNESIUM
- ELECTRIC STEEL
- FERROMANGANESE
- ELECTRIC PIG IRON
- ACETYLENE VIA ARC PROCESS

0 1 2 3 4 5 6 7 8 9 10 11 12
POWER COST (mills/kwhr)

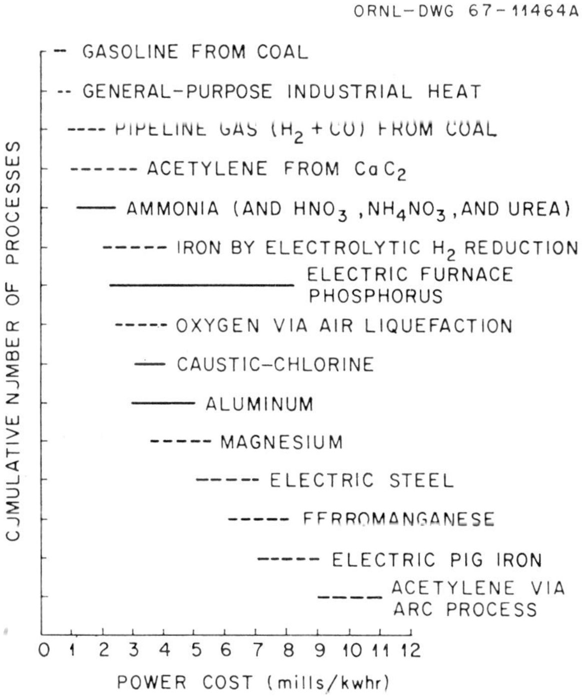

Figure 1 Elasticity of the Demand for Power.

LEGEND:

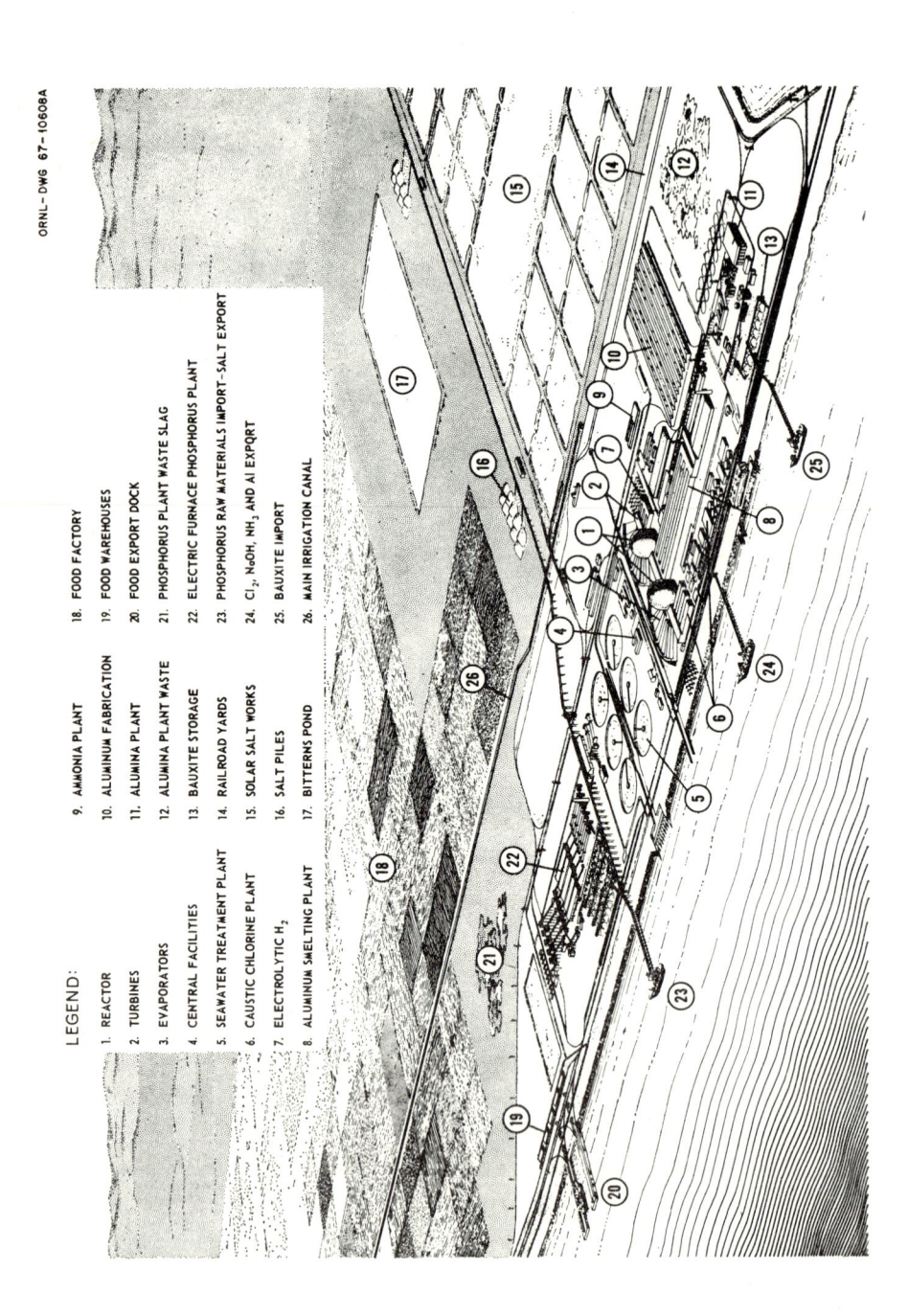

1. REACTOR
2. TURBINES
3. EVAPORATORS
4. CENTRAL FACILITIES
5. SEAWATER TREATMENT PLANT
6. CAUSTIC CHLORINE PLANT
7. ELECTROLYTIC H_2
8. ALUMINUM SMELTING PLANT
9. AMMONIA PLANT
10. ALUMINUM FABRICATION
11. ALUMINA PLANT
12. ALUMINA PLANT WASTE
13. BAUXITE STORAGE
14. RAILROAD YARDS
15. SOLAR SALT WORKS
16. SALT PILES
17. BITTERNS POND
18. FOOD FACTORY
19. FOOD WAREHOUSES
20. FOOD EXPORT DOCK
21. PHOSPHORUS PLANT WASTE SLAG
22. ELECTRIC FURNACE PHOSPHORUS PLANT
23. PHOSPHORUS RAW MATERIALS IMPORT-SALT EXPORT
24. Cl_2, NaOH, NH_3 AND Al EXPQRT
25. BAUXITE IMPORT
26. MAIN IRRIGATION CANAL

I HAVE WRITTEN a preface to this study on nuclear-powered industrial and agroindustrial complexes for several reasons:

First, I wish to stress the importance of the findings of the study. Combining the outputs of energy-intensive industrial processes, and clustering the plants around a nuclear reactor, is not a new idea. However, prior to this study, no really systematic analysis of such a complex had been made. Though the economics of such centers depends sensitively upon the prices that the industrial products can command, I find it most encouraging that even with fairly conservative assumptions substantial internal rates of return can be achieved in such nuclear-powered complexes.

Second, the study gives added incentive to the development of extremely low-cost energy sources. The demand for energy for chemical processing is decidedly elastic. If power can be produced at *much less* than the 3 mills or so per kilowatt-hour that TVA estimates for Browns Ferry No. 3, then we may see chemical processes increasingly substitute energy for other raw materials. To take an extreme example, power at 1 mill could play an important role in the liquefaction of coal. If these extremely low costs of energy are ever achieved, the demand for energy could be expected to rise dramatically, as indicated in a semiquantitative manner in the adjoining illustration. In a sense then, this study provides a strong incentive for the long-term development of the most advanced nuclear breeders—reactors that it is hoped can supply energy at much less than 3 mills/kwhr.

Third, though industrial complexes of the general type described here are not fundamentally new, the combination of these with highly rationalized agriculture based on desalted water is a new and very interesting idea. The relative emphasis to be placed on the agricultural and the industrial aspects of the energy center was a matter to which the study group gave serious thought. The balance which finally emerged represents a careful weighing of the views of the agricultural and industrial experts who participated in the work. In any case the agricultural and industrial elements of the study are

Figure 2 Nuclear-Powered Agro-Industrial Complex. An artist's conception of an agro-industrial complex stretching along the shore of a coastal desert. It produces and consumes 2000 Mw of electricity and 1,000,000,000 gal of fresh water per day, employing two large reactors for the energy source. It includes a 300,000-acre farm, irrigated with water from a seawater evaporator, and industrial plants to produce aluminium sheet and bar stock, electric furnace phosphorus, caustic-chlorine by brine electrolysis, and ammonia from electrolytic hydrogen. Associated facilities include a solar salt works, a railroad marshaling yard, an artificial harbor, and docks for import of raw materials and export of food and industrial products. Not shown are a town and other living quarters for about 100,000 persons.

well separated and documented so that those more interested in the one or the other, or in the combination, can readily find what they need.

A study such as this, with its rather general approach, is not intended to prove that a nuclear-powered energy complex in a specific location ought or ought not to be built. Such judgment must come only after a very detailed examination of a specific site that takes into account all local and regional economic and political factors. I am pleased that several specific site studies are now under way, and we hope that at least some of these detailed studies will lead to actual construction of nuclear-powered energy centers.

In conclusion, I want to thank all the people who worked so diligently in preparing the study. Particular thanks go to Professor E.A. Mason, who headed the study during his stay in Oak Ridge in the summer of 1967; to John Michel, Deputy Director of the study; to Commissioner James T. Ramey, who provided strong support for performing the study; and to R.P. Hammond, whose ideas have formed the basis for much of this study.

Introduction

In June 1967 the Oak Ridge National Laboratory started a study of the technical and economic feasibility of "nuclear-powered industrial and agro-industrial complexes", primarily as an avenue to industrial, agricultural, and general economic advancement in developing countries. Such a complex, shown schematically in Fig. 3, might consist of a large nuclear reactor station producing both electricity and desalted water. The electricity would be consumed in adjacent industrial processes and for pumping water, while the desalted water could be used either for municipal and industrial purposes in an industrial complex or in an irrigated agricultural complex located in a coastal desert region.

There are many different forms that energy-centered complexes can take. Possible complexes might include only the reactor coupled with an energy-consuming industry or with pumping stations for lifting and transporting groundwater to agricultural irrigation projects and for general industrial and urban use. An example of the latter case is described in a companion report* as applied in an irrigation scheme using pumped groundwater for the Ganges Plain in India.

---

* Perry R. Stout, *Potential Agricultural Production from Nuclear-Powered Agro-Industrial Complexes Designed for the Upper Indo-Gangetic Plain*, ORNL-4292 (to be published).

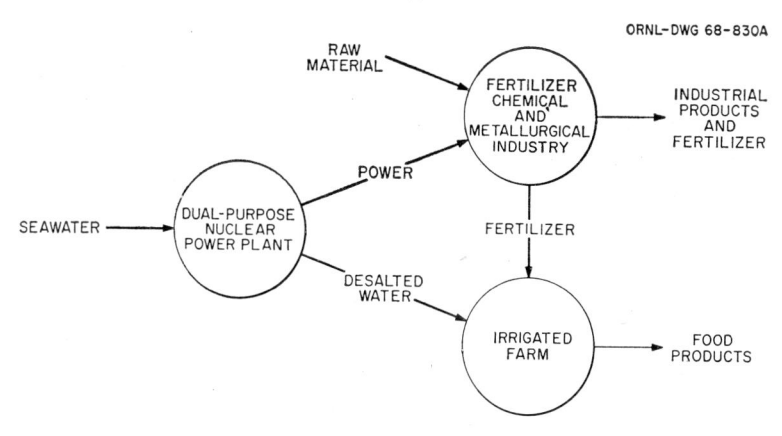

Figure 3 Agro-Industrial Complex.

The recent report of the President's Science Advisory Committee on *The World Food Problem* provides much of the motivation for the present study. This report* concludes, in part:

1. "The scale, severity, and duration of the world food problem are so great that a *massive, long-range, innovative* effort unprecedented in human history will be required to master it."

2. "Food supply is directly related to agricultural development and, in turn, agricultural development and *overall economic development* are critically interdependent in the hungry countries."

The principal question set by the ORNL study team was: How and to what extent could the low-cost energy anticipated from nuclear reactors be used effectively to increase both industrial and agricultural production, with particular attention being given to applications in developing countries?

Background for the study

A study of integrated nuclear agro-industrial complexes seemed appropriate at this time for several reasons. Starting in 1966 the nuclear reactor generating capacity sold to the utility industry in the United States had increased dramatically.[†] The cost of producing electricity from the largest of these reactors has been estimated to be less than the alternative costs for producing

---

* *The World Food Problem*, A Report of the President's Science Advisory Committee, The White House, May 1967.

† U.S. AEC, Division of Operation Analysis and Forecasting, *Forecast of Growth of Nuclear Power*, WASH-1084 (December 1967).

electricity from fossil fuels in many regions of the United States.[†,‡] Furthermore, developments now under way on advanced breeder reactors give prospects of further reductions in the costs of generating electricity.[§,‖,±] Electricity is already an important "raw material" in the production of many chemicals and metals,[**] and the future availability of such low-cost power is likely to increase its role.[††,± ±] A preference for power-intensive processes should lead to changes in the technology which will, in turn, affect the economics of the chemical and metallurgical industries and in some cases eliminate the dependency on certain key raw materials. The importance of this is magnified by the mobility of nuclear energy since a nuclear reactor, unlike a hydro plant or even an oil- or coal-fired plant, can be built "anywhere" without suffering a significant fuel cost penalty. These developments open the possibility of underdeveloped countries that now lack fossil fuels becoming self-sufficient in energy and then in many heavy chemicals, including the basic fertilizers.

Coupling large advanced nuclear reactors with seawater evaporators incorporating an improved heat transfer surface suggests that it may be feasible to use desalted seawater in irrigation agriculture.[*] In these dual-purpose plants, high-temperature steam from the reactor is used for production of electricity in a turbine-generator; the exhaust steam is then used as the heat source in a seawater evaporator for the production of fresh water. The projected cost of water from such plants, though much less than what has been demonstrated so far, still is higher than most irrigation farmers usually pay

---

† Tennessee Valley Authority, *Comparison of Coal-Fired and Nuclear Power Plants for the TVA System*, Chattanooga, Tenn., Office of Power, June 1966.

‡ G. L. Decker, W. B. Wilson, and W. B. Bigg, "Nuclear Energy for Industrial Heat and Power", *Chem. Eng. Progr.* **64**(3) (March 1968).

§ Appendices to *An Assessment of Large Nuclear Power Sea Water Distillation Plants*, Annex A, Interagency Task Force, Office of Science and Technology, March 1964.

‖ J. A. Lane, "Economics of Nuclear Power", *Ann. Rev. Nucl. Sci.* **16** (1966).

± T. D. Anderson et al., *Technical and Economic Evaluation and Four Concepts of Large Nuclear Steam Generators with Thermal Ratings Up to 10,000 Mw* (ORNL report to be published).

** J. M. Holmes and J. W. Ullmann, *Survey of Process Applications in a Desalination Complex*, ORNL-TM-1561 (October 1966).

†† R. E. Blanco et al., *An Economic Study of the Production of Ammonia Using Electricity from a Nuclear Desalination Reactor Complex*, ORNL-3882 (June 1966).

± ± Meyer Steinberg, *The Impact of Integrated Multipurpose Nuclear Plants on the Chemical and Metallurgical Proccess Industries. I. Electrochemonuclear Systems*, BNL-8754 (Dezember 1964).

* R. P. Hammond, *Desalted Water for Agriculture*, prepared for the International Conference on Water for Peace, paper No. P/384, May 23-31, 1967 (to be published).

although in many cases these prices are subsidized. It was recognized, however, that crop water requirements using distilled water may be less than generally had been believed to be the case. Its use in agriculture would nevertheless require intensive farm practices and skillful management.

Recent developments in both industrial and agricultural technologies further enhance the viability of such a complex. The electric furnace process for the production of phosphorus is of particular importance to developing countries that do not have sulfur,* especially in view of the recent rise in the price of sulfur.† Also of importance are the recent developments in water electrolysis, which could eliminate the need for natural gas or petroleum as a source for the hydrogen required in ammonia synthesis.‡ Inexpensive hydrogen will undoubtedly find other large industrial uses, such as reduction of iron ore to produce steel.

Recent advances have been made in agriculture, as evidenced by the new varieties of rust-resistant dwarf wheat and rice developed largely under the sponsorship of the Rockefeller and Ford Foundations. Under conditions of adequate fertilization and management, these varieties yield more than twice as much per acre as ordinary varieties. The water required to raise the grain needed to sustain an adult is much reduced by the use of these new crop types when coupled with efficient management practices.

These separate technologies, if judiciously combined, may provide developing countries a means of combating the imminent food shortages as well as providing a means of "leapfrogging" in their technological development. The advantages of combining these technologies into a single complex are twofold: first, the energy source can be larger than would otherwise be the case, and because of economics of scale the unit cost of power and therefore of each of the products is reduced; and second, byproducts or waste products from one process can serve as raw material for adjoining processes.

Industrial complexes somewhat like the ones described in this study are by now well known in the world. One of the best examples is the petrochemical SASOL complex near Johannesburg, South Africa; others are located in Trombay, India, and Texas City, Texas. The complexes described here differ from these in two respects: agriculture, based on desalted water, is part of some of the complexes studied in this report; and nuclear energy,

---

* Currently, phosphatic fertilizers are primarily produced by treating phosphate rock with sulfuric acid.

† T. V. O'Hanlon, "The Great Sulfur Rush", *Fortune* 77, 107 (March 1968).

‡ Allis-Chalmers, *Design Study of Hydrogen Production by Electrolysis*, ACSDS-0106643, vols. I and II (October 1966).

rather than coal or petroleum, is the fundamental raw material upon which these complexes are based. This existence of economically sound, integrated industrial complexes suggests to us that the idea of similar complexes based on nuclear energy is well worth serious further and detailed study.

Organization of the work

In approaching the study, it was decided to begin with a survey of the component parts of an agro-industrial complex. Lists were prepared of many industrial and agricultural products, and it was quickly realized that many eliminations and choices could be made and technical interrelationships uncovered without reference to a particular locality. On the other hand the availability of labor, materials (including suitable land), and markets for end products are strongly affected by the locale, so that a compromise between specific and general studies had to be made.

The study therefore proceeded along two parallel lines: first, "building block" information on industrial processes and farm crops was developed, and, secondly and simultaneously, information concerning the geography, demography, and economics of several coastal desert regions of the world was obtained. More specifically the work fell into the following categories.

1. The basis or rationale for the assumed costs of power, steam, and desalted water. This was divided into two time reference periods: cost ranges expected from plants using current reactor and evaporator technology, and cost ranges projected or anticipated from plants using advanced breeder reactors and advanced evaporator concepts.

2. The effects of the cost of electricity upon the technologies and total costs of various chemicals, fertilizers, and metals which require large amounts of electricity in their production. This work included studies of the effect of integrating a number of these energy-intensive processes into various industrial (nonagricultural) complexes which would be served jointly by a nuclear-powered generating station.

3. The effects of the cost of water on the total production costs of a variety of selected crops. This work entailed the development of water-yield relationships, quantities and costs for fertilizer, labor, seed, etc., and the capital costs involved in developing coastal desert regions for growing these crops under year-round intensively managed farming. While the cost per unit of production of agricultural products remained as the focal point, emphasis was also placed on obtaining the maximum productivity of water.

4. The economics of combining a nuclear electric generating station, an industrial complex, and an agricultural complex into an agro-industrial complex.

5. The geographic factors, such as topography, soils, climate, mineral resources, economic factors, and shipping costs, which would influence the nature and feasibility of nuclear-powered agro-industrial complexes in various parts of the world. This included a preliminary review of the social implications and possible problems of implementation in developing countries.

Reporting the results

The intent of this report is to describe the work performed, including a discussion of the rationale for the assumptions used, and to present the conclusions and recommendations for further work. Quantitative relationships have been included in the attached appendices to allow the reader to adjust the results for changes in assumptions in the manner of a "do-it-yourself kit". A separate summary report is being concurrently issued; more-detailed reports in several of the major subject areas will be published later, as follows:

Title of Report	Author	ORNL No.
1. Nuclear Energy Centers: Industrial and Agro-Industrial Complexes—Summary Report	Gale Young and J. W. Michel	1291
2. Potential Agricultural Production from Nuclear-Powered Agro-Industrial Complexes Designed for the Upper Indo-Gangetic Plain*	Perry Stout	4292
3. Data Obtained on Several Possible Locales for the Agro-Industrial Complex	T. Tamura and W. J. Young	4293
4. I. Steelmaking in an Agro-Industrial Complex II. Acetylene Production from Naphtha by Electric Arc and by Partial Combustion	A. M. Squires W. E. Lobo	4294
5. Problems in Implementation of an Agro-Industrial Complex	J. A. Ritchey	4295
6. Tables for Computing Manufacturing Costs of Industrial Products in an Agro-Industrial Complex	H. E. Goeller	4296

* Not prepared under auspices of the U.S. AEC but included in this series because of the close relationship to this project.

Perhaps it is desirable to mention what this study did not include. It was not intended to be a study *of* or *for* a particular country or region. Further, it could not be of sufficient depth to provide the basis for investment decisions; for example, no detailed market analyses or surveys of the adequacies of the countries' related infrastructures were conducted. In general, a *financial* analysis was not made nor was the nuclear-powered agro-industrial complex compared with other alternatives for achieving similar benefits. Finally, it should be recognized that the reason for not examining in this study an agriculture *only* complex based on desalted water was that single-purpose water-only plants have not yet been designed that will give water costs as low as those obtainable from the dual-purpose electricity/water plants.

Acknowledgments

During the summer of 1967 the Laboratory brought together a full-time study group staff of 16 engineers, economists, scientists, and agricultural experts under the direction of Professor E. A. Mason of the Massachusetts Institute of Technology. This staff was assisted by six consultants who worked on special topics, and by an advisory panel of 13 distinguished consultants from industry, government, and academic institutions. The panel met for three two-day review sessions during the summer. Experts from nine industrial organizations provided information concerning capital and operating costs for various industrial processes, while a large number of other contributors provided information on various other aspects of the project. The names and organizations of the participants are listed in Appendix 1 A, and we wish to acknowledge their help.

Summary

An intensive short-term study was made to evaluate the technical and economic feasibility of applying large nuclear energy centers for (1) the production of basic industrial products in the United States and in developing countries and (2) the production of both industrial and agricultural products using desalted water at coastal desert regions, primarily in developing countries of the world. This report describes the work performed in connection with this study, and the following summary section briefly discusses the most significant results in the main areas of work and presents the overall conclusions of the study. Detailed conclusions and recommendations are given in Chap. 9 of this report.

Two generalized models were used. In the first, the object was to determine the effect of various costs for electricity and water on the cost of production of industrial and agricultural products. Electricity and water were therefore considered to be purchased from outside the complex; the costs of the electricity, water, and raw materials required were varied parametrically over ranges selected to include conditions around the world.

In the second model, the object was to estimate the total investment, operating costs, income, and rate of return for integrated nuclear-powered industrial complexes. Since the electricity and water required for production uses would be produced within the complex, the costs of electricity and water in this second model were not estimated directly, but rather all the capital and operating costs for producing these inputs were included in arriving at the total costs for operating the overall complex under consideration.

In both models, various levels of production capacity were considered. Two sets of economic conditions were employed—one for conditions in the United States and one for developing countries. These conditions primarily consisted of assumed sets of costs of plant construction, raw materials, and labor, and the sale prices of finished products. Uniform methods were adopted to allow for interest during construction, depreciation, working capital, etc., using a range for the cost of money from 2.5 to 20%. No allowances were specifically made for taxes, nor were marketing expenses, including transportation costs, provided for generally. All costs and incomes were estimated at the 1967 level, with no allowance for escalation.

Three types of economic analyses were made to indicate the profitability of the concepts considered in this study:

1. For industrial products—the maximum cost of electricity which would give the same manufacturing cost as obtained by using an alternative non-energy-intensive process.

2. For industry or agriculture—the maximum power cost or water cost which would give a production cost equal to the current selling price.

3. For industry or agriculture or for complexes involving each or both—an internal rate of return which represents the cost of money at which the present value of the manufacturing cost, including investment, equals the present value of the income from product sales.

Sufficient information is presented to enable other forms of analyses to be performed so that comparisons with other possible investment opportunities may be made, but such comparisons were not a part of this study.

Power and water technology and cost bases

The technology and the associated cost of production of power from a nuclear reactor and the cost of desalted seawater from evaporator plants were established for reference use throughout this report. Two time periods were considered: (1) 10 years in the future (designated "near-term"), using somewhat improved current technology consisting of light-water reactors with multistage flash evaporators, and (2) 20 years off (designated "far-term"), using the advanced technology of breeder reactors and combination vertical-tube and multistage flash evaporators. Cost estimates of equipment and operating expenses were prepared for each time period and for various methods of financing. Table 1 summarizes the basic power costs for United States conditions which were developed and used in this study.

Table 1 Power Costs for Large Multiple Reactor Stations, 3880 Mw (electrical). (Power costs in mills per kilowatt-hour; load factor = 90%. Numbers in parentheses represent primarily operating costs of overhead and maintenance, insurance, and fuel cycle.)

Reactor Technology	Cost of Money			
	2.5%	5%	10%	20%
Near-term, light water	1.8 (1.2)	2.1 (1.3)	2.9 (1.4)	4.8 (1.6)
Far-term, advanced breeder	0.8 (0.2)	1.2 (0.3)	2.0 (0.5)	4.3 (1.1)

It should be recognized that these costs are illustrative estimates only and that, particularly for the far-term breeder reactor, the costs should be considered with uncertainty limits of at least ±20%. For example, increasing the capital cost of a breeder reactor by 25% would increase the power cost (at 10% cost of money) to 2.4 mills/kwhr; simultaneously lowering the load factor to 0.8 would increase the power cost an additional 0.4 mill/kwhr.

For dual-purpose plants producing power and desalted seawater, no cost allocation for the two products was attempted; but incremental costs of adding additional capacity for each were obtained for several sizes of plants and for costs of money from 2.5 to 20%. For the near-term technology the incremental power cost varied from 0.8 to 3.8 mills/kwhr (2.5 and 20%

respectively), and the incremental cost of water varied from 12 to 49¢ per 1000 gal (also at 2.5 and 20% respectively). The corresponding figures for the far-term case were 0.3 to 3.3 mills/kwhr and 5 to 34¢ per 1000 gal.

Use of power

A number of electricity-intensive processes were investigated to determine the effects of power cost on total manufacturing cost. This work involved the compilation of all the many cost components and their variation with the size of the production facility. Where possible, these processes were compared with a competing non-electricity-intensive process to determine the "break-even" power cost, that is, the cost of power at which the manufacturing costs by the two processes are equal. The two most important basic fertilizer materials, nitrogen (as ammonia) and phosphorus (as phosphoric acid), are in this category. Ammonia via water electrolysis was compared with ammonia via steam reforming of methane or naphtha, while phosphoric acid made by the electric furnace process was compared with phosphoric acid from the sulfuric acid acidulation of phosphate rock. Figure 4 shows these comparisons and illustrates the higher relative profitability of the electric furnace phosphorus process.

The manufacture of caustic and chlorine is normally done by brine electrolysis, and, unlike aluminum (see below), the raw material costs (salt) are usually quite low. For a cost of money of 10% and salt at $3 per ton, the production cost for chlorine from a 1000-ton/day plant (assuming no credit for the coproduced caustic and allowing for all capital charges including a 10% return on investment) is as follows:

Power Cost (mills/kwhr)	Manufactured Cost (dollars per ton of Cl_2)
2	33
4	40

For a chlorine selling price of $50/ton (recent U.S., f.o.b.), reducing the power cost from 4 to 2 mills would result in an appreciable increase of profits. In developing countries where the coproduct, caustic, is more in demand and sells for as much as $80/ton, the profitability is even greater. Caustic and chlorine (as hydrochloric acid) may be used either singly or together as a scale-preventative treatment for the seawater feed to an evaporator plant. Thus, electricity would in effect replace sulfur (as sulfuric acid) for treating seawater in evaporator plants.

106   *Desalting Seawater*

The manufacture of aluminum was also evaluated in some detail. Since an alternative process is not available for this product, a geographical comparison was made. For example, low-cost (2 mills/kwhr) power at a hydro site 6000 miles from the raw materials was compared with a nuclear-powered

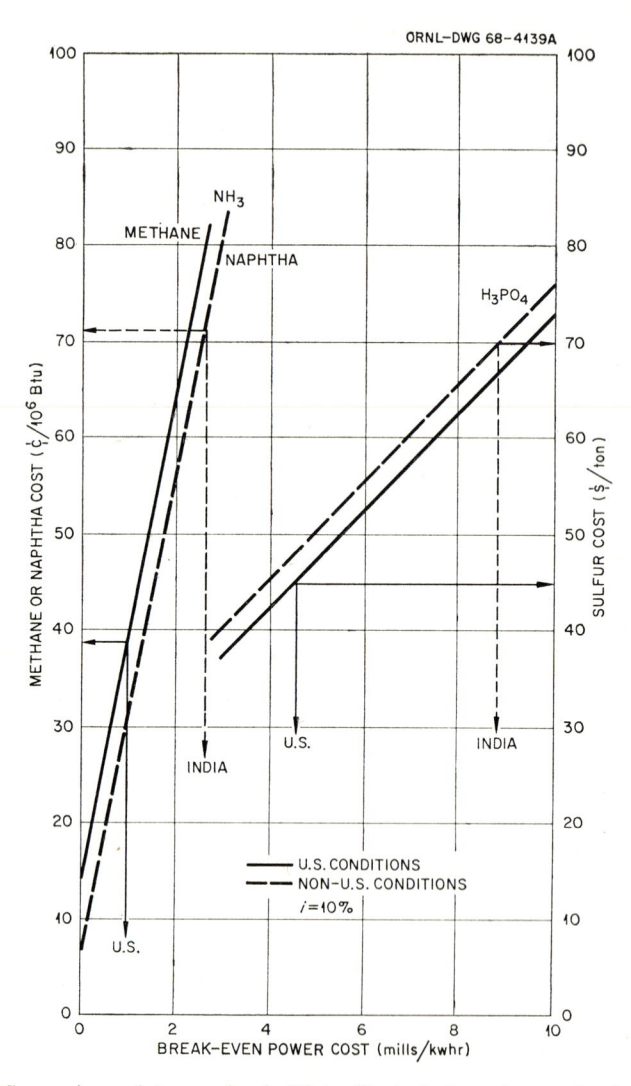

Figure 4 Comparison of Ammonia via Water Electrolysis vs Ammonia via Steam Reforming of Methane or Naphtha and Comparison of Phosphoric Acid by the Electric Furnace Process vs Phosphoric Acid by the Sulfuric Acid Acidulation of Phosphate Rock.

Nuclear Energy Centers

site 1000 miles from the bauxite source. For this case a "break-even" power cost range of from 2.5 to about 6 mills/kwhr was obtained for a wide range of parameter values (e.g., cost of money from 2.5 to 20%, plant capacity from 60 to 685 tons/day, and bauxite costs from $3 to $14/ton). In this comparison, imported alumina at $60 to $77/ton was assumed to be the raw material for the aluminum plant at the hydro site.

Other processes that were examined, but in less detail than those mentioned above, were (1) chemicals from evaporator discharge brine, including salt (NaCl), potassium chloride, and magnesium; (2) iron and steel by hydrogen reduction; (3) acetylene from naphtha (or methane) using the electric arc process; and (4) cement and sulfuric acid from gypsum (obtained from seawater).

The industrial complex (a group of interrelated industries without an on-site power plant) was also evaluated by the break-even power cost method. ["Break-even" in this connection denotes the cost of power at which production cost, including all capital charges (at a given cost of money, i) just equals income from the sale of products.] For a United States location with $i = 10\%$, the break-even power cost varied from about 4 to 6 mills/kwhr, depending on the product mix and the size of the complex. Two typical examples of the more than 70 cases evaluated are given in Table 2, indicating the effects of different product mixes, United States vs foreign location, and the influence of the cost of electricity on the attractiveness of the complex. The effect of the size of the complex and the cost of power on the rate of return is illustrated in Figure 5 for a particular product mix under United States conditions.

In general, the selection of processes studied was limited to those requiring relatively large amounts of electricity. These were primarily basic products which would usually be further processed before use or be used as raw materials for other processes. However, to test the effect of manufacturing secondary products on the break-even power cost, the manufacturing costs for three fertilizer materials—urea, ammonium nitrate, and nitric phosphate—all made from ammonia—were computed. The overall profitability was appreciably increased by including the manufacture of these products in a complex.

A generalized comparison was made of the relative profitability of a large integrated industrial complex [2500 Mw (electrical)] with an equivalent industry made up of small plants dispersed throughout the country near the market or point of consumption. This comparison thus indicated the tradeoff between the savings in manufacturing cost, due to the low-cost power from

Table 2 Two Typical 1000 Mw (electrical) Industrial Complexes.

	U.S. Complex		Non-U.S. Complex	
Products (tons/day):				
Ammonia—1500				
Phosphorus—560, as P_2O_5				
Aluminium—257				
Caustic-chlorine—500, of chlorine				
Total value of products (10^6 \$)	129		172	
Power cost (mills/kwhr)	2	6	2	6
Production cost*	118	150	128	165
Break-even power cost (mills/kwhr)	3.3		7.0	
Total capital investment (10^6 \$)	277		303	
Products (tons/day):				
Ammonia—1630				
Phosphorus—800, as P_2O_5				
Caustic-chlorine—355, of chlorine				
Total value of products (10^6 \$)	85		118	
Power cost (mills/kwhr)	2	6	2	6
Production cost (10^6 \$)*	71	106	107	130
Break-even power cost (mills/kwhr)	3.6		4.6	
Total capital investment (10^6 \$)	99		112	

* Capital charges computed for a 10% cost of money.

Nuclear Energy Centers

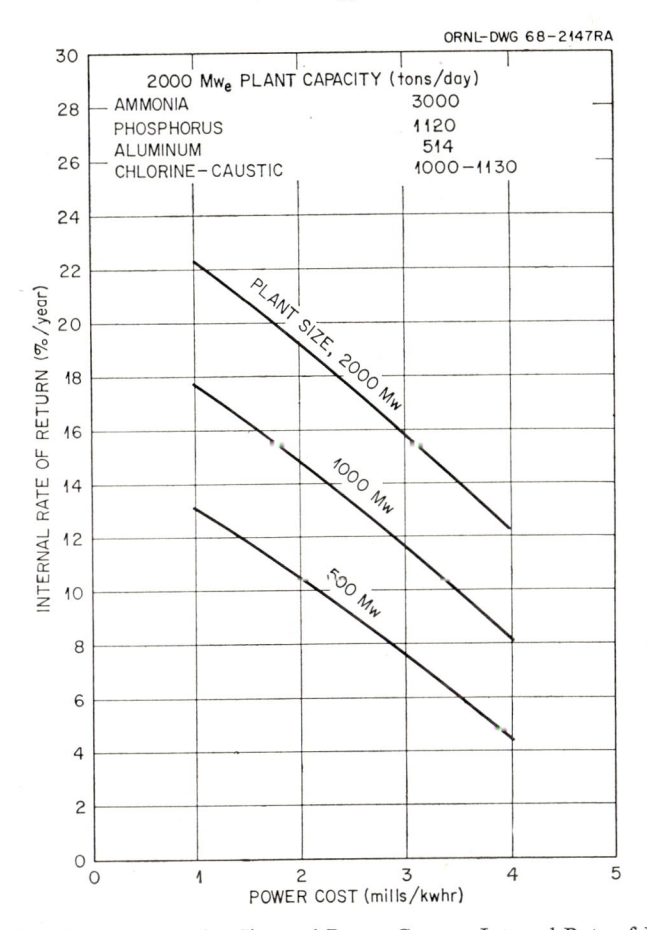

Figure 5 Effect of Complex Size and Power Cost on Internal Rate of Return.

a large captive power station along with size scaling and jointness advantages, and the increased transportation costs required to deliver the products to the markets. Table 3 summarizes this comparison for a non-U.S. case of shipping half the products from a large complex by rail 300 miles and half by sea 1000 miles and using a cost of money of 10%. Transportation costs for raw materials were not allowed for but would probably represent an additional advantage for the large complex at a seacoast location. As indicated in this table, a large complex could produce the same products (ammonia, phosphorus, aluminum, and caustic-chlorine) for two-thirds the investment and for considerably lower production cost. With the more probable value of 5 mills/kwhr for the small plant's power cost, over one-half

Table 3 Estimates of the Economic Advantages of a Large Integrated Industrial Complex.

	Power Cost (mills/kwhr)	Product Manufacturing Cost (dollars/year)	Product Transport Cost (dollars/year)	Income from Sales (dollars/year)	Investment in Industry (dollars)	Direct Return on Investment (%)
		$\times 10^6$	$\times 10^6$	$\times 10^6$	$\times 10^6$	
Large complex	3	264	25	462	640	27
Small plants	3	296		462	960	17
	5	341		462	960	13

Table 4 Comparison of Internal Rates of Return* for Several Nuclear-Industrial Complexes.

	Product Mix I†		Product Mix II†	
	U.S.	Non-U.S.	U.S.	Non-U.S.
Reactor Type				
Light water	11.4	16.1	13.1	16.3
Fast breeder	12.9	16.8	14.9	17.3
Thermal breeder	14.1	18.0	16.5	19.1
Size (Light-Water Reactor)				
500 Mw (electrical)	4.5	9.7		
1000 Mw (electrical)	7.4	12.7		
2000 Mw (electrical)	11.4‡	16.1		
Specific Location§				
1000 Mw (electrical) (LWR)	18.7§			

* The internal rate of return represents the maximum cost of money which may be used and just meet all expenses including return on investment, amortization, and interest during construction at this rate as well as the normal operating costs. Taxes and marketing expense are not included.

†	Product Mix I	Product Mix II
Ammonia, tons/day	3000	3080
Phosphorus, tons/day	1120	1500
Aluminum, tons/day	514	
Caustic-chlorine, tons/day	1130/1000	2260/2000
Power, Mw (electrical)	2048	2026

‡ Increasing the reactor cost from $125/kw to $150/kw would reduce the rate of return by 0.6. Eliminating the production of NH$_3$ and thus decreasing the power consumption to ~1000 Mw (electrical) increases the return by about 0.6.

§ Tailored product mix for a Florida location near phosphate rock deposit: 1180 tons/day of phosphorus and 685 tons/day of aluminum ingot.

of the difference in annual manufacturing costs may be attributed to the difference in the assumed power costs.

Several examples of a nuclear industrial complex were developed to illustrate the advantage of scaling the power source and jointly using other common facilities as well as using intermediate or waste products from one process by another. Rates of return were computed for a number of such complexes varying in size from 500 to 2100 Mw (electrical) for both United States and foreign conditions with different technologies and product mixes. As indicated in Table 4, rates of return varied from less than 5% to about 19%, the smallest value being obtained for a 500 Mw (electrical) United States case with current technology and the highest rate of return for the 2100 Mw (electrical) foreign advanced-technology case. Note the large increase in return for the 1000 Mw (electrical) LWR United States case when the product mix is tailored to a specific location.

Other uses of electric power which were included in some cases were: (1) power delivered by transmission lines to off-site load centers, (2) power used for pumping water within the evaporator plant and in the irrigation system, (3) auxiliary power for use within the complex, and (4) power for an associated town.

Use of water

The production of distilled water was considered only for agro-industrial complexes located in remote coastal desert regions where the water was used primarily for irrigation. While water for general urban use could be produced in an industrial-only complex, this was not specifically considered in this study.

Water used for irrigation supplied primarily the evapotranspiration requirements of the crops. These requirements were estimated for ten crops, including grains, vegetables, oil crops, fruits, and fiber. Crop yields and their response to varying levels of water application were also estimated by a review of the available data and consultation with many experts. The yields assumed are those regularly obtained today by the top 20% of farmers on good irrigated land. The values adopted are shown in Table 5. In the context of a highly mechanized and intensively managed farm, the direct costs for each crop were compiled. These included the costs for labor, fertilizer and chemicals, seeds, storage, market preparation, etc. Current prices paid to farmers were obtained for each crop so that relationships between the return (above direct costs) and the price of water to the farmer could be

Table 5 Crop Water-Yield Relationships

Crop Type	Crop	Water Use Inches	Water Use Gallons per Acre	Fertilizer Applied per Acre (lb) N	Fertilizer Applied per Acre (lb) P$_2$O$_5$	Yield (lb/acre)	Food Value (Cal/lb)	Efficiency of Water Use Yield (lb/gal)	Efficiency of Water Use Gallons per 2500 Cal
			×10^3			×10^3	×10^2		
Grain	Wheat	20	543	200	50	6.0	14.8	11.1	152
	Sorghum	27.6	749	150	80	8.0	15.1	10.7	154
Pulses	Peanuts	34.5	937	120	80	4.0	18.7	4.3	313
	Dry beans	20.6	559	70	70	3.0	15.4	5.4	302
Oil	Safflower	33.4	907	200	50	4.0	14.2	4.4	404
	Soybeans	33.4	907	100	50	3.6	18.3	4.0	343
Vegetables	Potatoes*	16	434	200	120†	48	2.79	111	81
	Tomatoes*	19	516	200	150	50	0.95	116	227
Citrus fruit	Oranges*	53.1	1442	180	30	44	1.31	30.5	628
Fiber	Cotton*	34.5	937	300	100	1.75 (lint) 2.8 (seed)		4.9 (total)	

* Due to marketing considerations, the acreages of these crops were restricted.
† 45 lb of K$_2$O was also applied.

obtained. Estimates were also made of the fixed costs, including the irrigation system, buildings, roads, equipment, and allowances for land reclamation and drainage facilities. It was then possible to estimate the relative profitability of producing various crops as a function of the price of water. For example, it was shown that some crops, such as tomatoes, citrus, and cotton, would have positive returns above direct costs with a water cost of 30¢ per 1000 gal or higher, while all other crops considered, with the exception of safflower, sorghum, and soybeans, could do so at 20¢/1000 gal.

The maximum water cost allowable so that the *total* production costs (direct plus fixed or capital expense) equal crop revenue, was obtained for several crops using a cost of money of 10%. For wheat this price was about 8¢/1000 gal, for peanuts, 12¢/1000 gal, and for potatoes greater than 35¢/1000 gal. These figures are quite sensitive to the assumed crop prices; for example, increasing the assumed price of wheat from 2.7¢/lb (paid to farmers in exporting countries) to 3.3¢/lb (delivered price to India) increases the maximum allowable cost for water to nearly 17¢/1000 gal.

Three types of cropping systems were evaluated: mixed crop, high profit, and high caloric production. All three obtained their irrigation water supply from a 1000-Mgd (million gallons per day) desalting plant at two levels of assumed water cost: 10 and 20¢/1000 gal. Table 6 summarizes these evaluations, and Fig. 6 indicates the effects of changes in the cost of water and in the crop price levels on the rate of return.

The two sets of crop prices used in Fig. 6 are (1) those paid to farmers in exporting countries, to cover the case of entering world markets, and (2) an import price, to cover the case of internal consumption of the food. The use of set 2, which was 30% above set 1, significantly increased the profitability of the farm.

Another vital assumption made is that suitable crop varieties for the region will be available, which in some cases implies development programs including experimental farms and involving years of advance effort. In general, the uncertainties associated with agriculture appear somewhat greater than for the industrial enterprises.

Integration of power and water production and uses

Combining the nuclear reactor, turbine-generator, evaporator, industry, and farm into one large enterprise, a nuclear agro-industrial complex, necessitated the development of a physical model and an economic model. The physical model, partially depicted in the frontispiece, is based on use of a relativ-

Nuclear Energy Centers

ely flat coastal desert region and includes provision for all the required facilities to operate the complex including a town and small, family farm plots (not shown) for the farm employees and some of the industrial workers. This model also includes the required facilities for storage and shipping of all raw materials and products.

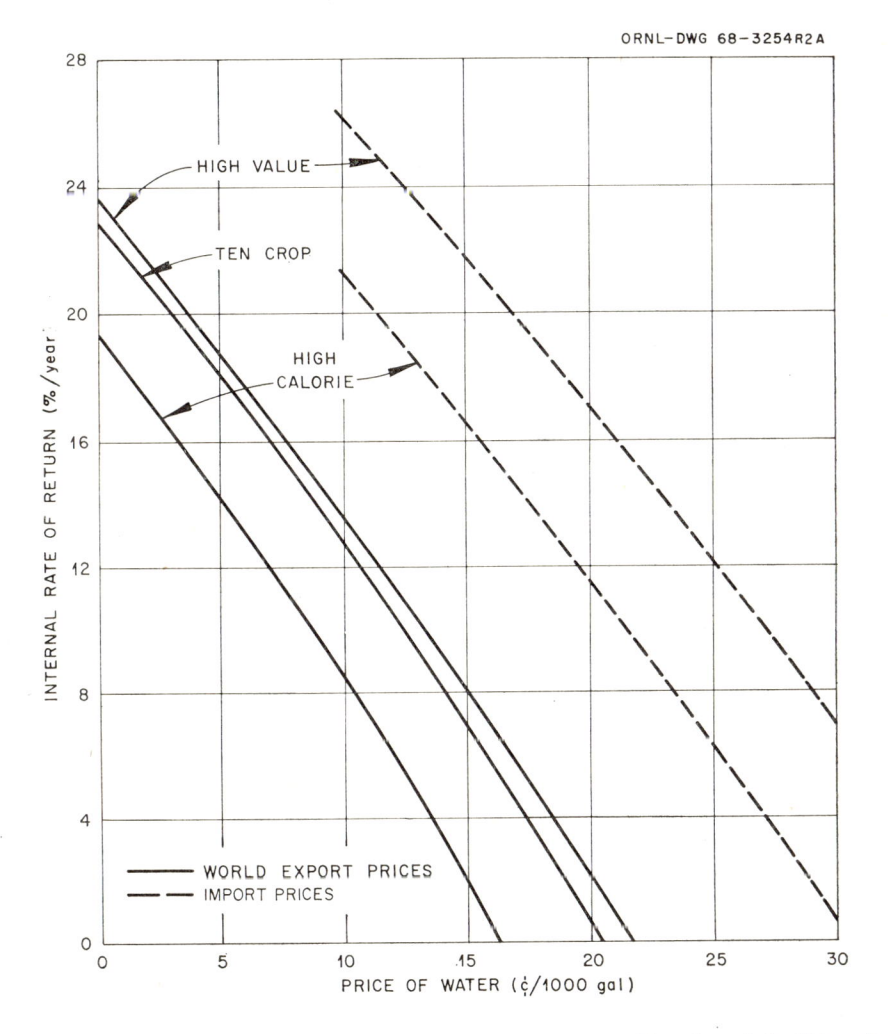

Figure 6 Internal Rate of Return for the Three Farms as a Function of the Price of Water.

Table 6 Summary of Farm Systems.

	Ten Crops	High Value	High Calorie
Farm size, thousands of acres	280	320	300
Percentage of water temporarily stored	18	26	24
Production			
Millions of tons per year	3.6	3.1	3.3
Billions of Calories per year	4080	4800	5680
Investment, millions of dollars	285	306	295
Operating costs, millions of dollars per year, at:			
10¢/1000 gal	115	102	92
20¢/1000 gal	148	135	125
Gross receipts at import prices, millions of dollars per year	206	195	182
Internal rate of return,* %, at:			
10¢/1000 gal	25	26	21
20¢/1000 gal	16	17	12
Millions of persons fed†	4.5	5.3	6.3
Protein per person fed, g/day	91	107	79
Water used per person fed, gpd	200	170	145
Investment per person fed, dollars	66	58	47
Operating cost per person fed, ¢/day, at:			
10¢/1000 gal	7.0	5.3	4.0
20¢/1000 gal	9.0	7.0	5.4

* Including all operating and overhead expenses, allowances for interest during construction, and all capital charges.
† 2500 Cal/day.

The economic analysis of the complex consisted in the itemization of capital expenditures, operating costs, and receipts from the sale of products. This was done for two levels of reactor/evaporator technology at several sizes for both United States and foreign cases. The internal rate of return was computed as a figure of merit for each case. Table 7 shows a condensed version of one of the many economic evaluations made and illustrates the difference between the application of near-term (light-water reactor and multistage flash evaporator) and far-term (advanced breeder reactor and vertical-tube evaporator) technologies for a non-U.S. location.

Table 7 Comparison of Technologies for Nuclear-Powered Agro-Industrial Complexes Producing Aluminium (685 Tons/Day), Ammonia (1740 Tons/Day), Phosphorus (765 Tons/Day), Caustic-Chlorine (1500 Tons/Day), and Food Industrial power, 1585 Mw (electrical)

	Near-Term Technology		Far-Term Technology		
	Primary Products	Secondary Products, NH$_3$ Converted to Urea and Ammonium Nitrate*	Primary Products plus Grid†	Primary Products; No Grid; Steam Bypass‡	Primary Products; No Industry; Steam Bypass§
Station size, Mw (thermal) (two reactors)	11,100		11,900	10,800	8820
Net electrical power, Mw (electrical)	2100		2900	1935	312
Desalted water, Mgd	1000		1000		
Farm size, acres	320,000		320,000		
Investment, millions of dollars					
Nuclear island	166		261	246	217
Turbine-generator plant	120		118	83	20
Grid tie	0		13	0	0
Evaporator plant	403		279		
Industrial complex	570	730	570		0
Farm	306		306		
Working capital	79	85	71	65	28
Harbor	35		35		30
Town	32		32		14
Fuel inventory	70		191	174	141
Total±	1781	1947	1876	1790	1035
Annual operating costs, millions of dollars					
Power and water plant	47		18	16	21
Industrial complex	133	152	133		
Farm	56		56		62
Total	236	255	207	205	83
Annual sales, millions of dollars					
Fissile material	7		16	15	11
Grid	0		20	0	0
Industrial products‖	347	407	347		0
Farm products‖	194		194		
Total	548	608	577	556	205
Income minus expenses, millions of dollars	312	353	370	351	122
Internal rate of return, %/year	14.6	16.1	16.5	16.4	10.1

* Only changes in numbers are listed; all other numbers are the same as listed under Primary Products.

† Due to higher initial steam conditions than obtained with the near-term case, more electricity is made; excess [∼1000 Mw (electrical)] is sold to a grid at 3 mills/kwhr.

‡ Evaporator operated using some bypass of prime steam to achieve full water output of 1000 Mgd with no excess (grid) power produced.

§ Only sufficient power is generated to operate reactor, evaporator, and farm; 85% steam bypass.

± Interest charges during construction not included in this total but are allowed for in computing the internal rate of return.

‖ Import price level used.

This table also illustrates the effects of several of the variables considered on the internal rate of return: (1) the effect of manufacturing secondary products (i.e., ammonium nitrate, urea, etc.) improves the return and (2) in the far-term case, bypassing 25% of the prime steam directly to the desalting plant and thereby reducing the electricity generation does not appreciably affect the return. Bypassing about 85% of the steam to provide only enough power to operate the desalting plant and the farm (no industrial power) decreased the internal rate of return for the far-term complex from 16.5 to 10.1% and for near-term technology from 14.6 to 7.4%.

For non-U.S. conditions the effects of size, industrial product mix, crop price level, and assumed capital and operating costs were varied to determine the sensitivity of the internal rate of return to variation in these parameters. Increasing the size of the complex from 525 Mw (electrical) industry/320 Mgd farm to 2100 Mw (electrical)/1280 Mgd gives an increase in the internal rate of return of about four percentage points. Eliminating the production of aluminum while increasing the production of caustic/chlorine and phosphorus decreased the internal rate of return by about one point. The use of export or world price levels for both the industrial and farm products decreased the return by about six points. Table 8 summarizes the sensitivity analysis by giving the amount of change required in the pertinent cost and income assumptions to cause a one percentage point change in the internal rate of return. Finally, the incremental rates of return for the addition of the food factory to nuclear-industrial complexes varied from 7 to 15%.

Table 8 Sensitivity Analysis: Changes Required to Give a One Percentage Point Change in the Internal Rate of Return for the Far-Term Complex with Grid

	Amount of Increment (dollars) $\times 10^6$	Percentage Change in Estimate
Nuclear island cost	102	39
Evaporator cost	112	40
Industrial complex cost	108	19
Farm cost	121	40
Operating expenses	21	10
Sales		
Industrial products	21	6
Farm products	21	11

Applications

Five coastal desert regions around the world were studied as potential areas for the location of an agro-industrial complex. These were located in India (Kutch), southeastern Mediterranean (Sinai), Baja California, Peru (Sechura), and Western Australia. The individual localities were studied to test the sensitivity of the many assumptions made in relation to actual conditions existing in the world so that the breadth of the applicability of the agro-industrial complex could be estimated. The main locale parameters investigated were climate, soils, topography, physical resources, and transport facilities. In general, irrigation agriculture on the scale envisaged appeared feasible at all five locales, although more detailed information would be required before a realistic evaluation could be made. Also better resource surveys and market analyses would be required.

There appeared to be many possibilities for industrial applications both in the United States and overseas, particularly near large deposits of phosphate rock or bauxite.

A preliminary study of implementation problems as influenced by the social, political, and cultural environment was made. This study resulted mainly in the definition of a number of potential problem areas, and although some recommendations were made, no specific plan for implementation was developed.

Overall conclusions

The main overall conclusions derived from this study project may be listed as follows:

1. Significant economic advantages appear possible by coupling an industrial complex with a large nuclear heat source as compared with equivalent (same producing capacity) dispersed smaller industry. The advantages are generally greater in developing countries than in the United States but are highly sensitive to the product mix selected and to local conditions which affect the cost of raw materials or other manufacturing costs. Industrial complexes based on a capacity requirement as small as 500 Mw (electrical) in some circumstances give internal rates of return of 10% or more.

The effect of advances in nuclear power technology, that is, use of breeder reactors, would significantly improve the internal rate of return. In the most striking example, the case of a particular nuclear-industrial complex in the United States, the substitution of a breeder reactor for a light-water reactor increased the rate of return from 13.1 to 16.5%, a 25% increase in the internal rate of return.

While this study did not consider nonnuclear energy sources, there may be some situations where fossil-fuel or hydro sources are preferred. In general, the concept of an integrated industrial or agro-industrial energy center is not dependent on the type of energy used, although the inherent characteristic of relative freedom of location is an important advantage for nuclear energy.

2. The use of coastal desert regions for producing a variety of agricultural products by irrigation with desalted water appears technically feasible and generally competitive with food produced on existing farms. The extra cost for the expensive water is at least partially offset by the opportunity to conduct intensive year-round food-factory agriculture in favorable growing climates with many conditions under unusually good control. It appears that using year-round cropping patterns that might be employed on actual farms, the calorie requirement for a man can be met using less than 200 gal of water per day. For the high-calorie cropping pattern which also satisfies the minimum protein requirements,* sufficient food (2500 Cal/day) for one person could be *produced* for abour 3¢/day with an initial investment of about $165 per person. These numbers are based on the incremental costs of adding an evaporator desalting plant and farm to a large agro-industrial complex. There were also several nonmonetary advantages identified for such a food factory located in coastal desert regions, for example: (1) the reliability of food production is increased since more of the production variables would be under control; (2) freedom from, in many cases, restrictive traditions or cultural practices so that the economic advantage of large-scale mechanized farming can be realized; (3) the internal requirement for an on-going agricultural research program could be expanded to benefit the country as a whole—the food factory could become a center for education, training, and research to also improve the conventional agriculture; and (4) unused or "waste" land could be made productive and valuable.

The food produced in off-site conventional agriculture which can be attributed to the application of the fertilizer made in the complex but surplus to the food factory requirements† could provide the minimum diet for 60 to 90 million people. The investment cost attributable to the required fertilizer production facilities, including the appropriate portion of the nuclear power

---

* Quantity of proteins is adequade but not the quality or the required protein spectrum.

† Up to 95% of the ammonia and 98% of the phosphorus produced would be shipped from the complex.

plant (LWR), would be about $7 to $4 per person fed, and the operating cost would be 0.5 to 0.3¢ per person per day. The range for the number of persons fed is based on the range of the expected increase in grain yield per pound of applied plant nutrient as discussed in ref. 2 of Chap. 1 and assumes some simultaneous improvements in production practices.

3. Though it appeared that the above two conclusions were generally valid at the five locales studied, a much more detailed analysis of a locale using specific local data, including the prevailing financial costs, would be required before specific implementation of such a project could be attempted. This would include, in addition to actual soil, mineral resource, climatological, and labor surveys, a detailed marketing and logistic analysis as well as consideration of the many socio-political implications. Finally, the alternatives which may be available for achieving similar benefits would need to be evaluated to establish the best approach for actual implementation.

Reference

1. Oak Ridge National Laboratory, operated by Union Carbide Corporation for the U.S. Atomic Energy Commission, November 1968. ORNL-4290, UC-80-Reactor Technology.
   Report available from Clearinghouse for Federal Scientific and Technical Information, National Bureau of Standards, U.S. Department of Commerce, Springfield, Virginia 22151.
   Reprint includes chart, Prefatory Note, Introduction, and Summary.

IX

PROSPECTS FOR DESALTED WATER COSTS[1]

William E. Hoehn

Rand Corporation, Santa Monica, California

Introduction

AS A PART OF the RAND Corporation's Middle East Project, sponsored by the Ford Foundation, I have been asked to prepare a "definitive" report on the cost of desalted water from large desalting plants. That this paper does not satisfy that request can be attributed to two factors. First, neither I nor RAND presently have the technical expertise, resources, and time necessary to develop the requisite data base, methodology, and models required to perform parametric analyses across the broad spectrum of possible locations, outputs, costs, interest rates, fuel prices, etc. Second, and more fundamentally, even if those resources were forthcoming, there is no reason to believe that the results of such a study would be in any sense "definitive". This arises because of the total lack of data on the construction, operation, and maintenance costs of large desalting plants—there is not one existing plant against which such paper studies can be checked. There is none in operation, there is none under construction,* there is none for which comprehensive firm bids from component suppliers have been obtained. There is nothing to draw on but an immense collection of (largely well-intended) paper studies of greater and lesser relevance to estimation of deslating costs. This fundamental caveat must be borne in mind in evaluating the estimates of desalting costs in this (or any other) paper. Let us consider some of the estimating difficulties.

A major source of difficulty in the estimation and discussion of desalted water costs arises from the lack of acceptance of a single method of calculating water "costs". That is, in order to produce desalted water at "useful" prices, it is generally accepted that the plant should be designed as a dual-purpose

---

* With the exception of one in the Soviet Union for which no economic data has been provided.

plant, producing both electric power and desalted water. The desalting process is limited as to the maximum acceptable temperature of steam used in the brine heating stages, and this maximum temperature is well below the temperature at which steam is most efficiently produced in nuclear reactors or fossil-fueled boilers. Consequently, high pressure, high temperature reactor or boiler steam is first utilized in a turbine—leading to electric power production—and is extracted at lower temperature and pressure for use in the desalting plant.* As a consequence of this dual-purpose operation, certain costs will be incurred that cannot be attributed solely to production of one or the other product, so that rules must be specified as to how such joints costs are to be allocated to the two products.

Different rules will, of course, lead to different proportional allocations of joint costs, and, thus, to different "costs" for water and for power. A recent study sponsored by the International Atomic Energy Agency (IAEA)† discusses no less than nine different allocation methods, which lead to nine different estimates of desalted water "costs", and correspondingly, to nine different electric power "costs". (Since total costs are fixed, any allocation method resulting in lower water "costs" must of course result in higher power "costs".) The highest water "cost" of the nine was half again as expensive as the lowest water "cost",§ which suggests the range of variability that may be present in any listing or comparison of desalting costs from various sources, in which the cost methodologies are not specified in detail. Nor, indeed, is the IAEA listing exhaustive; other rationales can be (and have been) employed to shift relative production "costs" one direction or the other.

Once a methodology is explicitly or implicitly agreed upon, one can turn to estimates of investment and annual operating costs for large dual-purpose plants. As noted previously, there is no direct experience on plants of this type from which to start. In the context of the Middle East,‡ there is not

---

* The temperature and pressure at which it is extracted are still above the lower limits of the turbine-generator set, however, so that the desalting process is not utilizing steam that would otherwise be "wasted".

† *Costing Methods for Nuclear Desalination*, IAEA, Technical Report Series No. 69, Vienna (*1966*).

§ *Ibid*. Water "costs" varied from 25.9 to 40.0 cents per thousand U.S. gallons for their particular reference plant, while power "costs" ranged inversely from 4.90 to 1.25 mills per kilowatt-hour.

‡ Throughout, by Middle East, I refer generally to the area east of the Nile and south of Turkey and Iran, but the major focus of much of the paper will include only the Eastern Mediterranean area.

even good data on the cost of the steam source, as the size of the heat source for even fairly small dual-purpose plants is well in excess of any existing plant in that region. Thus, one is confronted with the necessity to make use of historical U.S. costs and estimate the differential, if any, for various Middle East locations. The same applies to the power-generating equipment, with the added proviso that data on back-pressure turbine costs (and operational problems) is more limited than that for conventional turbines.

As for the desalting plant, the primary interest is in desalting capacities for agricultural purposes rather than for more limited municipal or industrial uses, so that the scale of interest is in the high tens or hundreds of millions of gallons* per day (MGD). The largest desalting plants presently in operation (using fossil-fueled heat sources) are on the order of 5–6 MGD, so that the scale-up envisioned is considerable.† In addition, performance parameters have been pushed to higher levels in the scaled-up desalting plants, raising some feasibility/reliability questions. Thus, few of the components of estimates of desalting plant costs can be traced back to experiences at existing plants.

Finally, one must specify interest and/or fixed charge rates, annual operating, maintenance, and replacement costs, etc., to calculate the cost of desalted water. Not only do questions of mechanical reliability and design efficiency enter in here, but also one must consider possible future trends in inputs over the operating life of the plant—possible changes in items such as nuclear fuel cycle costs, fossil fuel prices, annual plant operating factor, etc.

We shall want to consider the effects of many such permutations of assumptions, not so much to try to establish absolute "costs" as to understand which assumptions can markedly reduce or increase water costs. The absolute level of water "costs" will only be of secondary concern as an indication of the extent to which desalting offers some promise for widespread agricultural uses. As this is a question of the price at which water is provided (and its marginal value in agricultural production) as well as the "cost" of desalting, we shall be content to simply give rough estimates of the extent to which desalted water could be made available at various prices.

I propose to proceed along the following lines. First to be examined are the two proposed projects for which the most detailed evaluations have

---

* Throughout this paper, "gallons" denotes U.S. gallons.

† Although not a direct function, as the larger plants generally envision using several parallel modules of smaller capacity; thus a 150 MGD plant might use three parallel 50 MGD modules, or about an order of magnitude scale-up.

been made—a proposed plant in Southern California* and a proposed plant in Israel. The intent is to try to develop from these detailed studies some bench-marks around which some less-detailed parametric studies can be placed to permit an evaluation of costs under a wide variety of assumptions. Finally, we shall consider the potential impact of some grandiose proposals for massive-scale desalting and industrial development.

As the above paragraphs suggest, this report is not intended to be either a comprehensive study of desalting costs or a detailed survey of the desalting literature. The former is too ambitious, while, as a consequence of substantial cost escalations of major components since the 1965–1966 time period, the latter is too unrewarding.

Costing methodologies

The problem of cost allocation for dual-purpose plants arises because not all of the investment and operating costs can be uniquely attributed as incurred solely on behalf of one or the other of the two products. Of these jointly incurred items, the most significant and costly is the steam source; steam is utilized in both power and water production. Moreover, certain economies of scale occur in dual-purpose facilities relative to single-purpose plants. That is, total annual costs for a dual-purpose plant will, in general, be less than the sum of annual costs of a single-purpose power plant and a single-purpose water plant, each with capacity equal to the outputs of the dual-purpose plant. In principle, these benefits of scale can be used to reduce the cost of water or power or both below the cost of production from equivalent-capacity single-purpose plants. Therefore, rules have to be specified to allocate jointly incurred costs (and, thereby, benefits of scale) to either power or water production before water (and power) costs can be determined.†

An IAEA study has identified and explored nine such allocation rules of general applicability in calculating desalting costs.§ While a discussion of the merits of all of these would not be productive, a brief description of each

---

* Although recent disclosures of substantial and unexplained increases in cost estimates limit the usefulness of this case.

† If, of course, one is engaged in deciding between two prospective plants with equal outputs of both products under the same groundrules, the issue is simpler. The one with the lower total annual cost is preferable, and the issue of power and water costs need not arise.

§ *Tech. Report 69, op. cit.*

may help to explain the potential diversity of water costs, and the pitfalls in the methodologies.

The IAEA report distinguishes between *proration* methods and *credit* methods. Proration methods divide total costs of the dual-purpose plant into two parts proportional to external or internal single-purpose costs. Credit methods, as the name implies, impute costs to one or the other of the products by reference to external single-purpose costs, and attribute the remainder of the dual-purpose total costs to the other product.

Proration methods

Three proration methods are discussed:

Method A   Proration on the basis of single-purpose water and power plants.
Given two single-purpose plants of equal capacity to the capacities of the dual purpose plant, allocate dual-purpose total costs to water and power in proportion to the total cost of each product from the single-purpose plant.*

Method B   Proration on the basis of power generated.
This method allocates to the cost of water the cost of the added amount of power that could have been produced had the dual-purpose plant heat source been used exclusively for power production rather than for power and water production. The significance of this measure is not readily apparent.

Method C   Proration on the basis of available energy (exergy).
Available energy is a thermodynamic function giving the amount of energy that can be transformed into mechanical work in accordance with the Second Law of thermodynamics. Basically, all costs are allocated to water and to power in proportion to their consumption of heat energy in the steam produced. It has been a favored method of those who approach desalting from the engineering or technical side and are bemused or befuddled by the joint production allocation methods of economists.†

Credit methods

It is safe to say that credit methods have been more frequently applied (and misapplied) than have proration methods, if only because once a credit amount has been determined, the calculation of costs is vastly simplified.

---

* One might also use this proration factor to allocate only the dual-purpose *joint* costs, but the rationale for this procedure is less clearcut.

† Needless to say, there is an inverse relationship here; most economists are bemused and/or befuddled by the engineering arguments, not to mention such abstruse concepts as Mollier diagrams, entropy and enthalpy, turbine efficiencies, etc.

The IAEA report identifies six variants of credit methods.

Method D Generation cost of single-purpose power station of same net output as dual-purpose net power capacity.

The assignment of a "cost" per kilowatt-hour to power produced in the dual purpose plant allows a portion of dual-purpose total costs to be charged to power production, with the residual cost allocated to water production. This method allocates to water production almost all of the benefits of scale in joint production.

Method E Generation cost of a single-purpose power station of net output equivalent to that potentially available from the heat-source of the dual-purpose plant.

Here, instead of using the single purpose plant cost equivalent to the dual-purpose net power rating, the credit used is generating cost for a plant equal in size to that which the dual-purpose plant would produce if all of the heat energy were used for power production. The rationale for this seems somewhat obscure.

Method E' The cost of the added power that could have been produced had all the heat energy of the dual-purpose plant been used for power production is charged to water production.

The rationale for this seems even more obscure.

Method F Purchase price of the power paid by the purchasing utility.

This method ought not to be considered appropriate, and is listed only because of its inclusion in the IAEA study. This method confuses and commingles costs and revenues, with the results that the water "cost" so defined is *not* a cost, but rather a *breakeven price*, given the power price, at which water must be sold to fully cover total plant costs. Indeed, if a sufficiently high price for electric power can be obtained from a consumer (utility or otherwise), the so-called "cost" of water under this method could be zero or even negative.* Not that there is anything wrong with examining the relationship of breakeven prices—indeed, one has to consider revenues as well as costs in deciding whether to build the plant. But prices are *not* costs, and costs are the issue of this paper. As will be seen below, this method has been extensively and subtly misused.

Method G The cost of water produced in a single-purpose water plant of the same net capacity as the dual-purpose plant.

---

* For an illustration of negative water price using this method see my *Economics of Nuclear Reactors for Power and Desalting*, The RAND Corporation, RM-5227-1-PR/ISA, November 1967. Section X.

The obverse of method D, this method assigns to power production almost all of the benefits of scale in joint production, and generally leads to the highest water "cost" of any method considered. For this reason it is seldom used.

Method H   The cost of water from the least cost alternative source of supply in the absence of the dual-purpose plant, regardless of capacity.
This is a variant of method G, equally simple in principle, but the definition of the "least-cost" increment in practice is quite complex.

Which of these to use? The unfortunate fact remains that even after the more outlandish are discarded, there remain several reasonable allocation rules, and one cannot on strictly economic grounds marshal a compelling case for the elimination of all but one method.

On a pragmatic basis, however, it would seem that one of these indeed enjoys a clear preference. In the past, however intense the invocation of impartial economic principles in the introductory remarks, all of the specific non-parametric studies have used allocation methods that minimize the reported "cost" of water. One can thus conclude that the interests of all parties to those studies have been to award (at least) all of the benefits of economies of scale to the production of water. On this basis, method D can be considered to the be preferred method.*

Thus, we will endeavor to present all of the estimates of desalting costs in terms of method D, although most detailed studies have used variants of method F, and one easy-to-use parametric study to be discussed, modified, and utilized below uses method E (a slight variant of D). Method D seems more appropriate than method E—which assumes a lower power credit corresponding to a larger power output—when one considers the economies of scale in dual-purpose plants. An effect of Method D is to specify that the water plant is charged only for the incremental costs of increasing the heat source from the size required to produce the desired electric power, to the size sufficient to produce the desired quantities of both power and water. Now, the curves of cost versus capacity for nuclear steam systems† show falling cost with increasing capacity throughout the range, but declining more slowly at larger capacities (approaching an asymptote). Therefore, for a given increase in nuclear steam system rating, corresponding to a given

---

* Actually, method F, with a utility purchase price well in excess of equivalent single-purpose power costs (thus pushing water costs even below that which method D produces) is the undisputed favorite. However, gentle reader, you and I aren't about to use so economically unsound a method, are we?

†The remarks in this paragraph apply equally to fossil fueled steam sources.

water plant size, incremental costs of the increase in nuclear steam supply (reactor size) will be smaller the larger the rated electric power capacity. Therefore, the tendency to try to minimize water costs means that the electric power output of a desalting plant will generally be specified to be as large as can practicably be absorbed by the electric power grid in the region around the dual-purpose plant. But if this is the case, then method E is irrelevant, since the added power production assumed in determining its power credit could not usefully be absorbed.

There are some further considerations insofar as the utilization of method D is concerned. The first issue to be resolved concerns the specification of the alternative single-purpose power station of equivalent net output. In particular, this does *not* imply that the credit shall be based on the cost of a single-purpose plant of the *same* type. The credit is to be based on the least cost alternative single-purpose plant, evaluated under the same ground rules (interest rates, load factor, fuel cost, etc.) as apply to the dual purpose plant.

The identification of the least-cost alternative is no simple matter. There are at least three possible alternatives: nuclear fueled, fossil fueled, and hydroelectric plants. Although one should properly consider hydroelectric possibilities in addition to fossil fueled and nuclear fueled plants, hydro power is so location-dependent that, for our purposes, it may be dismissed from further consideration.*

Even limiting the alternatives to fossil and nuclear fueled heat sources, the comparison of least-cost alternatives is tricky. This arises because, for equal generating capacities, nuclear plants tend to have higher investment costs than fossil fueled plants, while nuclear plant annual operating charges (largely fuel cost) tend to be nearly independent of location and tend to be lower than comparable fossil fuel costs unless the plant location is very near major sources of fossil fuel production (coal deposits or oil or gas fields). What this suggests it that for a given plant capacity and a given fossil fuel cost, the lower the interest (or fixed charge) rate, the greater the relative advantage of nuclear power, and, conversely, the higher the rate, the greater the relative advantage to fossil fueled alternatives. In the following sections, we will generally be dealing with a range of interest and fixed charge rates, since alternative methods of financing desalting plants can lead to substan-

---

* Although I claim no special familiarity with the Middle East, the only hydro proposal I can recall in the eastern Mediterranean area was one to lift water from the Mediterranean over the intervening hills into the Dead Sea; the purpose of this was largely to stabilize the (lowering) water level in the Dead Sea, with power generation incidentally feasible due to the difference in elevation.

tially different fixed charge rates. Thus, in general, we will need to consider the generating cost of both fossil fueled and nuclear fueled single-purpose alternatives, with the power credit for use in estimating the dual-purpose plant's water costs based on the lower of the two alternative plants' generating costs.

Of course, this comparison of nuclear and fossil alternatives should reflect not only near-term fuel costs, operating costs, and capacity factors, but estimated future trends in these variables as well. This in turn raises some new problems regarding which of several methods should be used to average varying costs over time to reflect a single average cost. In some cases, rather than try to project likely costs over the 25–35 year lifetime, only costs over the early years are used, a procedure that is somewhat akin to discounting costs. This truncation procedure simples the calculation and could be considered to reflect a relatively high rate of time discount (although the rate of interest used in most such calculations suggests that others have markedly different time preferences).

In any case, the calculations are difficult and beset with wide uncertainties. Take, for instance, future uranium prices. If one looks at some sources (such as forecasts of nuclear growth rates or comparisons of fossil and nuclear plants), one would conclude that no more than a moderate increase in price would be experted; however, if one looks at other sources (such as studies in support of advanced reactor concepts), one would conclude that increasing uranium prices cause present-generation reactors to become non-competitive within a few years.

In general, the costs estimated herein are based in general on limited rates of change over time of the major variables, in keeping with general trends in most of the fuel and mineral markets over time. Alternatively, this could be viewed as the use of early-years results only with the future trends discounted to zero. In most of the other studies noted herein, future prices are either not considered or are based on similarly limited trends, so the results should be generally consistent.

One further point is deserving of mention. The power credits herein, unless otherwise indicated, are based on the use of capacity factors in the alternative selected equal to that assumed for the power portion of the dual-purpose plant. This, in effect, credits the dual-purpose plant with the same operational flexibility as a single-purpose plant. When considered as individual units, it may be reasonable to assume the two to enjoy the same flexibility, but from a utility system standpoint, this is less accurate.

During the course of the day (week, season, year) demand for power fluctuates requiring the addition or removal of generating capacity. From the standpoint of minimum-cost power generation to meet demand, the investment cost of existing capacity is irrelevant, so the decision as to which unit to add to or remove from service is based on incremental operating costs. That is, one first loads the unit for which operating costs are lowest, than the next-lowest-cost unit, etc., with the highest-cost units used only sparingly, to meet the peak daily or seasonal demands. This is generally called "minimum cost" or "order-of-merit" scheduling.

Now, on a utility system, because of technological improvements to new units and wear and tear on older units, newer units tend to have lower incremental operating costs than do older units. Thus, a unit, over its useful lifetime, can expect to be relatively heavily used during its early years, but used more and more sparingly in later years, as newer capacity additions display lower incremental costs. As a result, the annual operating factor tends to decrease over time, and the average capacity factor over the plant lifetime is lower than the early years capacity factor.

Consider now the dual-purpose plant. By the method of calculating power credits, the lifetime capacity factor is assumed to stay at the high early-years level—that is, the plant is presumed to operate baseloaded throughout its lifetime. Reduction in powerplant load leads to inefficiencies in stem production and to higher dual-purpose plant costs,* but if the plant load factor is kept high throughout, departures from "order-of-merit" scheduling can occur, leading to higher utility system costs. Either way, the outcome is higher costs to the system (although not necessarily to the plant), so that an adjustment to the power credit would seem appropriate. However, since this involves evaluation of the entire power system to which the dual-purpose plant is to be added, this correction is seldom mentioned and almost never estimated. Although we mention it here, we will also not attempt to estimate an appropriate reduction to the power credit.

---

* Since water costs are based on power credit methods, fewer kilowatt-hours per year generated means higher water costs as well.

Some case studies in problems of cost estimation: the MWD plant*

As of a year ago, two proposed dual-purpose plant projects had been evaluated extensively for a particular location and under specific ground rules. One of these was a plant to built off Southern California on an artificial island, with the Metropolitan Water District of Southern California (MWD) to be owner of the water plant (with a limited interest in power generated), and with a group of three utilities (two private, one municipal) owning (most of) the power generating facilities. The other project was a cooperative U.S.–Israel effort proposing a plant to be built on the Mediterranean coast of Israel with U.S. financial assistance.

The MWD plant was to produce 1600 MWe power, and ultimately, 150 MGD of water; the Israel plant was initially (1965) to produce 200 MWe and 100 MGD. The Israel plant was subsequently uprated (in 1967 to 300 MWe and 100 MGD, with a new optimization and cost estimation, and was again revised to reflect cost increases as of June 1968. The MWD plant was re-evaluated in early 1968, but only the broad results have been made public.† Those broad results differ so extensively from the earlier study that this case is of little use as a checkpoint.

Another study of dual-purpose plants for Mexico's west coast reportedly is in the final completion stages, but detailed data is not yet publicly available.

I have elsewhere presented a detailed evaluation of the economics of both the MWD and the Israel desalting plants as originally reported, and with my own estimates of costs as of mid-1967.§ This re-estimation as of mid-1967 was necessitated by rapid and substantial escalation of nuclear plant and equipment prices, and of construction and installation costs, beginning in 1966 and extending so far as can be inferred to the present moment. These escalations have far outstripped the few percent rise in Wholesale or Con-

---

* As the article was in preparation for publication, some additional information was made available in a paper prepared for a symposium on desalting held in Madrid. See *Nuclear Industry*, January 1968, pp. 19–23. The data and informatiom reported therein do not appear to materially alter the thrust of the analysis in this section.

† See, e.g., *Los Angeles Times,* July 25, 1968, p. 3; *Wall Street Journal,* July 26, 1968, p. 8; *Nucleonics Week,* August 1, 1968; *Nuclear Industry,* August 1968, September 1968. This last reference reports that the power capacity from the MWD plant had been increased by about 400 MWe.

§ That is, predating the MWD re-evaluation, and both of the Israel revisions. See RM-5257, *op. cit.*, Section X.

sumer Price indices, and have sharply reduced the usefulness of most parametric studies conducted (or using inputs developed) during the period. As an example, three issues of General Electric's U.S. nuclear pricing handbook, dated September 1965, August 1966, and November 1966, listed the price for a 600 MWe nuclear steam supply system at, respectively, $22.6, $24.9, and $29.3 million, an escalation on these components of about 30 percent in less than 15 months.

The rapid upward trend in prices paid continued through much of 1967, with some slowing tendency discernable late in the year. Prices have reportedly continued to rise during much of 1968, but apparently are not now rising at a significantly more rapid rate than other equipment and construction indices (which, in turn, generally been increasing more rapidly than general price indices).* The widespread increases in construction costs and in major component prices (such as copper) have also raised desalting plant costs, although not as drastically as the steam supply costs. All of these factors contribute to the hazards of cost estimation.

An extreme example of cost increases is the case of the ill-fated MWD plant. The case is instructive both for the magnitude of the overall misestimation, and because the initial water cost estimates from this plant have been so widely quoted and used as a basis for extrapolation of future water costs. Unfortunately, not enough detailed information has been made available publicly to attribute escalation and/or gross underestimation by major components.

The original investment cost estimate for this twin nuclear reactor powered, 1600 MWe, 150 MGD dual-purpose plant was $444 million in 1965 prices. This figure included investment costs for water and power conveyance from the plant location on an artificial island one-half mile off the mainland to distribution points inland. The cost of water at the plant boundary was estimated to be 21.9 cents per thousand gallons† (and 27 cents per thousand gallons at an inland blending plant), based on 90 percent plant capacity factors, 30 year plant lifetime, 3.5 percent cost of money, and 5.44 percent fixed charge rate.

My own analysis of the MWD study in mid-1967, applying a correction for an assumed power credit in excess of production cost (method F) on MWD's entitlement to 100 MWe of the net power capacity (181.5 MWe

---

* My own estimates (*ibid.*, Section IV; Table 9, Fig. 2, *idem.*) are presently in need of escalation of some 20–30 percent to correspond to recently announced estimated costs.

† For those more familiar with water pricing in units of $/acre-ft., one cent per thousand gallons is about $3.25 per acre-foot.

gross entitlement), and with allowance for escalation and less optimistic assumptions of some other parameters,* led me to conclude

> ... that far from the widely quoted figure of 22 cents per thousand gallons, the production cost of desalted water from the MWD plant can scarcely be less than 30 cents per thousand gallons even with considerable optimism, and that a less optimistic evaluation of several important elements entering the cost calculations would suggest true production costs nearer 35 than 30 cents per thousand gallons.†

Following the decision by all parties to proced, bids were received from many major equipment items of the MWD plant evidently due to substantially higher bids than were estimated, a re-evaluation was undertaken. On the basis of that re-evaluation, the estimated investment cost of the plant (in spring 1968 prices) was given as $765 million, or 72 percent above the estimated 1965 costs. The cost of water at the plant boundary was then estimated to be 35 cents per thousand gallons§ (computed on the same basis as the earlier estimate of 22 cents per thousand gallons), an increase of 60 percent over the 1965 figure.

The breakdown of investment costs between MWD and the utilities was along the following lines. In the 1965 estimates, the utilities' share was $257 million and MWD's share was $126 million. The remaining $61 million (of the $444 million) was made up of federal contributions of $46 million by the Interior Department's Office of Saline Water (OSW) intended to offset the added cost of the desalting plant relative to the supposed cost level for later plants on a volume basis; the other $15 million was from the AEC and was intended to offset the cost of expected design interface problems between the nuclear steam supply system and the desalting plant.

In the 1968 estimates, the $61 million federal contribution was unchanged, the utilities' share had increased from $257 to $487 million (almost 90 percent), and MWD's share had increased from $126 million to $217 million (about 72 percent).

In late July 1968, a decision was reached not to proceed with the plant under the multiple ownership agreement then in effect. It has been suggested that Southern California Edison Company was the prime mover in the deci-

---

* The most optimistic of which surely is the assumption of 90 percent capacity factors for both power and water over the lifetime (30 years) of the plant. Most experience with U.S. nuclear and desalting plants would appear to support capacity factors on the order of 60–75 percent. It takes a small act of faith even to accept the "customary" 80 percent capacity factor for nuclear plants.

† RM-5227, *op. cit.*, p. 187.

§ The coincidence is fortuitous, but it helps make my track record look good.

sion not to proceed.* It was announced that AEC and Interior would continue to explore other alternatives to the specific plant, and it appears that several alternative mainland locations were explored.† All of these were much less favorably located for integration of the desalting plant into MWD's distribution system, however, and the latest (and possibly final) word on this project is that MWD has voted to "proceed" with the plant on the artificial island, but "not to start construction until mid-1970s,"§ with initial capacity of 50 MGD to be in operation "not later than 1980".‡ However, one report of the cost of water on a "go-it-alone" basis (without Federal subsidies) estimates costs of 91.7 cents per thousand gallons for the initial 50 MGD plant, and 56.2 cents per thousand gallons if capacity were increased to 150 MGD.‖

Thus the actual effect of the maneuverings and analyses of this project over the last year is to render estimation of water costs from this project subject to overwhelming uncertainties. One can suggest from the above references that actual costs would probably lie somewhere in the range from 35 to 95 cents per thousand gallons, but one could offer that accurate a forecast without reference to the details of this project.

One must also bear in mind that the proposed MWD plans is atypical in several respects of the plants we might wish to consider for the Middle East. First, owing to siting problems in the Southern California area, the least-cost siting solution was an artificial island to be built off the coast. So stringent (and expensive) siting constraints do not appear to arise in the case of the Middle East. Second, the MWD plant as initially configured had a net power capacity of 1600 MWe, or roughly equal to the total installed capacity of Israel, Syria, Lebanon, and Jordan. So large a total block of power (and such large unit sizes) could not be adsorbed in that area, unless demand for power were to be substantially increased. Yet, as we have already mentioned, water costs are quite sensitive to economies of scale, and, as power output increases,

---

* *Los Angeles Times*, July 25, 1968; *Wall Street Journal*, July 26, 1968. This is reasonable, since Edison's fixed charge rate would be substantially higher than that of the Los Angeles Department of Water and Power (a municipal utility) or the MWD (a regional authority with broad taxing power and tax-exempt financing). Thus, escalation would affect it most sharply of the three.

† *Nucleonics Week*, December 5, 1968, pp. 2-3.

§ *Los Angeles Times*, December 11, 1968, Part II, p. 1. The article is optimistically headlined "Nuclear-Fueled Desalting Plant Gets Go-Ahead".

‡ *Ibid.*

‖ *Nucleonics Week*, December 5, 1968, p. 3; no details of the basis for this estimate were given, other than that it would be on an artificial island. Presumably, since these estimates are so high, net power production would be lower than the original 1600 MWe.

water costs decline even for fixed water output. It is tempting but undemonstrable to argue that these two factors (expensive siting; cheap power from economies of scale) would tend largely to offset one another.

We also should note in passing that the original estimate of 22 cents per thousand gallons was calculated using a quite low fixed charge rate (5.44 percent) corresponding to a 3.5 percent cost of money (to MWD). Without wishing to belabor from the standpoint of theoretical economics the issue of "appropriate" fixed charge rates for efficient allocation of resources, one should note that the desalting process is highly capital intensive. Therefore, estimates of water costs will be strongly influenced by the fixed charge (and/or interest) rate assumed. Finally, the estimated water costs were based on an assumed plant life (for costing and depreciation purposes) of 30 years, and on an annual operating factor of 90 percent of capacity. The justification for such high values is quite speculative,* and failure to achieve either would lead to sharply higher actual water production costs.

What lessons can be drawn from this case? The loudest message appears to be a reiteration of the all too familiar rule: "Don't trust the cost estimates." No matter how one juggles possible breakdowns of investment costs, even accepting one report that the final cost estimates included an increase in the power capacity of 400 MWe,† it is not possible to attribute the increase of $321 million solely to conventional measures of escalation of material and equipment prices (plus the capacity increase)—in the final analysis some costs must have been incurred for items not included in the initial estimates. That is, the initial report must have failed in some respect to fully identify all costs for construction of the plant.

A supporting bit of evidence for this hypothesis would appear to be a remark last June by a senior vice-president of Southern California Edison Company (one of the participants in the MWD project):

> There are many kinds of lies. There are harmless exaggerations to make a point; there are little white lies; there are bald-faced lies; there are out-and-out whoppers. And at the very end of the spectrum of lies, there are atomic costs.§

Since this would have been some weeks after Southern California Edison received the re-estimate of costs for the MWD plant, it seems reasonable to infer that the initial cost estimate was the target he had in mind.

---

* See my RM-5227, *op. cit.*, for further comment and elaboration of these points.

† *Nuclear Industry*, September 1968, pp. 14–15; since confirmed by other sources.

§ W. R. Gould, speech to National Coal Association convention, June 1968; quoted in *Nuclear Industry*, August 1968, p. 15.

However, the latest estimates are just that—revisions of earlier estimates based on revised equipment prices and on a somehow enlarged and expanded purchase list. It is not by any means a "construction cost". Typical experience these days is for realized construction costs to outrun estimated costs, the result of a widespread estimation bias. Whether missiles or moon-rockets, airplanes or rifles, the military regularly encounters this phenomenon, and has yet to learn the right kind of compensation mechanism.* Similarly, and closer to home, the history of irrigation projects, dams, and other water works shows the same consistent pattern of biases: approval of project based on estimated pre-construction costs and benefits, followed by discovery that actual costs substantially exceed estimates, followed over time by the discovery that benefits fail to accrue as forecast, followed some years later by the inevitable revision of payout schedules and terms. Yet this estimation bias is rarely raised in such hearings—the Bureau of Reclamation (to name only one such agency) has certainly one of the worst records for accuracy of estimates, yet as it presents the economic analysis of project after project, no question is raised of the confidence that can be placed in the analysis.

Even closer to home, there is substantial evidence presently accumulating to suggest that the forecast investment costs of the large nuclear powerplants ordered before 1965 have been systematically underestimated (both costs to the manufacturer and to the purchasing utility). General Electric Company has already publicly admitted that it seriously underestimated the costs on a number of its early nuclear plant sales,† and several utilities have suggested that their share of costs are turning out to be higher than anticipated.§ As only a few of these plants have even reached initial operation, and data is still fragmentary, it is not yet possible to estimate a reasonable "bias factor" for use on dual-purpose plant cost estimates, but it would be surprising if one were not needed.

---

* The latest case seems to be the C-5A Galaxy transport, which has presently overrun cost estimates by 50 percent, with an ultimate overrun of 100 percent "possible". See, e.g., *Aviation Week and Space Technology*, **89,** No. 21 (November 18, 1968), p. 29. It seems a moot point whether the extrapolation of present jets to the C-5 is greater than the technical extrapolation to large-scale desalting plants.

† *General Electric Company 1966 Annual Report*, New York, 1967, p. 10.

§ See, e.g., *Nuclear Industry*, September 1968, p. 11; *Nuclear Industry*, May 1968, p.22; *Nuclear Industry*, February 1968.

The Israel plant

This plant, proposed for a location on the Mediterranean coast near the town of Ashdod, has been extensively studied and revised, but a lack of agreement over terms on which U.S. financial assistance would be made available has kept this project at the planning stage. The original study (by Kaiser Engineers) of plant rated at 200 MWe and 100 MGD was published in January 1966, with cost estimates based on 1965 prices.* Since financial terms had not been established, this report (and all subsequent Kaiser reports) presented parametric costs for fixed charge rates of 5, 7, and 10 percent. Plant optimization, where a function of fixed charge rates or interest rates, appears to have been carried out for the seven percent rate.

Delay in approval of plans together with a lengthening of nuclear plant lead-times† led in 1967 to the conclusion that the earliest possible plant start-up-date would be mid-1974. By that date, the Israel power network was believed to be capable of absorbing a 300 MWe block of power, rather than the 200 MWe originally planned. Accordingly, Kaiser Engineers prepared a new report based on capacities of 300 MWe and 100 MGD in terms of mid-1967 prices.§

In mid-1968, Kaiser Engineers published a further report for a plant of the same 300 MWe and 100 MGD capacities, presenting plant costs reflecting mid-1968 equipment prices. In addition, it projected some cost estimates including escalation to the plant completion date in 1975.‡

Table 1 gives the estimated desalted water costs at the plant boundary for each of the several reports prepared by Kaiser. These costs, however, are based on slightly differing assumptions and are not all directly comparable. All reports are based on the use of a utility-specified power credit that is invariant to changes in the fixed charge rate. Therefore, the method used is

---

* *Engineering Feasibility and Economic Study for Dual-Purpose Electric Power Water Desalting Plant for Israel,* Kaiser Engineers (in association with Catalytic Construction Co.), Report No. 66-1-RE, January 1966; hereafter cited as *1966 Kaiser Report.*

† Hoehn, RM-5227, *op. cit.,* see Section IV.

§ *Effects on Costs of Increasing Capacity to 300 Megawatts,* Kaiser Engineers (in association with Catalytic Construction Co.) Report No. 67-35-RE, December 1967; hereafter cited as *1967 Kaiser Report.*

‡ *Estimates Based on Mid-1968 Costs and Escalation Based on Plant Design Commencing Mid-1969 With Construction Completion 1975,* Kaiser Engineers (in association with Catalytic Construction Co.), Report No. 68-25-RE, June 1968, hereafter cited as *1968 Kaiser Report.*

Table 1 Cost of water from dual purpose Israeli plant, cents/thousand gallons

Plant size and year of estimated prices	Fixed charge rate		
	5%	7%	10%
200 MWe, 100 MGD; 1965 prices	28.6	43.4	67.0
300 MWe, 100 MGD; 1967 prices	23.6	40.2	66.0
200 MWe, 100 MGD; 1968 prices[a]	25.0	42.0	68.0
300 MWe, 100 MGD; 1968 prices[a] with 7% capital cost escalation to start up	37.0	60.0	97.0

[a] Two digit accuracy only.

Sources: *1966 Kaiser Report, 1967 Kaiser Report, 1968 Kaiser Report, op. cit.*

method F, rather than the more conservative method D. The interest components of the fixed charge rates of 5, 7, and 10 percent are, respectively, 1.6, 4.6 and 8.4 percent, with a 30 year sinking fund and 0.8 percent included in the fixed charge rate for insurance and taxes.

I have elsewhere discussed and will not further belabor the *1966 Kaiser Report*, other than to mention that, based on prices as of mid-1967, with a power credit more nearly reflecting method D than method F,* my estimate of the cost of water at the lowest (5 percent) fixed charge rate was then about 35 cents per thousand gallons, compared to Kaiser's 28.6 cents.† This 35 cents estimate is not comparable to any of the other figures in Table 1, as it uses a revised power credit, and is based on 200 MWe, rather than the 300 MWe in the 1967 and subsequent Kaiser estimates. My report was published prior to the *1967 Kaiser Report*, and so does not deal with the increased power capacity. Also, reports in 1967 suggested that the Government of Israel recognized that the plant would not be an economically justified proposition at the higher fixed charge rates studied.§ Therefore, no attempt was made in my report to adjust cost estimates for other fixed charge rates.

Given the larger-capacity plant proposal and the new estimated costs, the first adjustment we might wish to undertake would be to apply revised power credits corresponding to method D.

---

* The careful reader will note that I actually calculated the credit on the basis of the cost of a single-purpose power plant equivalent in capacity to that implied by the dual-purpose heat source (method E).

† Hoehn, RM-5227, *op. cit.*, Section X.

§ See, e.g., *Nucleonics Week*, May 26, 1966, p. 2.

140 *Desalting Seawater*

The exact nature of the alternative plant having generating costs of 5 m/kwh (or 5.3 m/kwh in the *1966 Kaiser Report*) was finally parenthetically revealed in the *1968 Kaiser Report*. In a single comment on the basis for escalation of cost estimates to mid-1968, the report notes:

> The power credit is based on the estimated cost of electrical generation from a single purpose power plant of comparable capacity *burning oil* in Israel.*

This brings us up against one of the previously-noted difficulties of using method D: the fact that increases in the heat-source size in going from a single purpose power plant to a dual purpose plant with the same power production may cause a shift from nuclear to conventional (or vice versa) if the crossover occurs between the sizes. The *Kaiser Reports* have throughout affirmed that nuclear was preferable to fossil-fueled heat sources for the dual-purpose plant.† This leaves the question whether nuclear or fossil-fueled plants would be preferred for a single-purpose power plant of 300 MWe capacity.

I have elsewhere presented a comparison of 300 MWe nuclear and coal-fired power plant generating costs based on plant and equipment costs in the U.S. as of mid-1967.§ That report estimated nuclear generating costs for a 300 MW plant in the U.S. at a 10.5 percent fixed charge rate at 5.33 m/kwh. Given allowances for escalation since mid-1967 to either of Kaiser's study dates and for the increased cost of shipping equipment to and constructing the plant in Israel, there is no way that nuclear generating costs for to small a plant could approach 5 m/kwh at a 10 percent fixed charge rate.

The problems of estimating differential costs for projects in foreign locations are many and difficult, and there is little data from which to work. An earlier reconstruction of costs for the Israel plant in comparison to those in U.S. arrived at an estimate of about $5 million for overseas costs, calculated as a residual, not on an item-by-item basis.‡ Also presented was a more detailed listing prepared by General Electric for a Congressional Committee investigating the high cost of the Tarapur, India nuclear project.‖ That listing recorded some $17 million in added expenses relative to a U.S. location

---

* *1968 Kaiser Report*, p. IV-4; italics added.

† Although not enough data in support of their assertion has found its way into the reports to permit verification or manipulation of that part of the analysis. One must take that conclusion on faith.

§ Hoehn, RM-5227 *op. cit.*, Table 15 and Section IV.

‡ *Ibid.*, Section X.

‖ *Ibid.*, p. 165–166.

for two twin-unit 200 MWe nuclear plants. However, the name of the game in that proceeding was to explain an apparently large discrepancy between Tarapur costs and some later bids in the U.S., so one should view G.E.'s enumeration of differential costs cautiously.

Starting from the estimated cost for a 300 MWe plant in the U.S. as of mid-1967, about $63 million,* added costs due to escalation and foreign location of perhaps 25 percent, plus a few percent for the cost of a "nonstandard" size, implies total costs on the order of $80–$85 million for a 300 MWe nuclear plant in Israel in late 1967 or early 1968. If we associate the low end of this cost range with the lower fixed charge rate and the upper end with the higher fixed charge rate,† we should have a reasonably conservative set of investment cost figures with which to work. (The higher the investment cost, the higher the power credit, and the lower the water cost.) These capital costs figures lead to the following rough estimates for the power credit (assuming the same 85 percent capacity factor) at each fixed charge rate: 3.7 m/kwh at 5 percent; 4.6 m/kwh at 7 percent and 6.0 m/kwh at 10 percent.§ Taking the same 85 percent capacity factor assumed in the Kaiser Reports, and without consideration of any deductions for operation outside of "order of merit" scheduling the plant lifetime, the revisions to the water costs per thousand gallons in the *1967 Kaiser Report* are as follows: 33 cents instead of 22.6 cents at 5 percent; 43 cents instead of 40.2 cents at 7 percent, and 60 cents instead of 66.6 cents at 10 percent.‡ Note that the effect of this computation is to raise the water cost at the lower fixed charge rates while lowering the water cost at the higher. When the consider fossil-fueled alternatives, the credit at 10 percent will revert to the 5 m/kwh used in the Kaiser Reports, and the power credit at 7 percent may also be lower; however, at 5 percent fixed charges, the nuclear plant should dominate. That is, if money is really available at 1.6 percent interest, so that one can achieve a 5 percent fixed charge rate, then one can produce power very cheaply from a single-purpose *nuclear* power plant, which has a higher proportion of costs charged to capital accounts. On the other hand, if fixed charge rates are

---

* *Ibid.*, Table 15.

† To reflect the variation in interest during construction due to variations in the capital charge rate.

§ Assuming $10 million average investment in nuclear fuel, 1.2 m/kWh fuel cost and 0.65 m/kWh for operation, maintenance, and nuclear insurance.

‡ Needless to say, neither my figures nor Kaiser's are accurate to the first decimal place—all of us are hoping that the first *digit* is accurate. Nonetheless, calculations and sliderules do give "answers" to several "significant" figures.

higher, reflecting higher interest rates, then the cost of power from a fossil-fired single-purpose plant becomes more attractive since the initial investment is so much smaller.

Let us now consider power credits when the single-purpose power-plant for comparison under method D is fossil-fueled rather than nuclear fueled. The Kaiser Reports make plain that the 300 MWe alternative from which their power credit is derived is evaluated at 10 percent fixed charges,[*] uses oil fuel[†] at 33 cents per million BTUs,[§] and generates power at 85 percent capacity factor at 5 m/kwh.

The fuel cost at a heat rate of 8913 BTU/kwh[‡] amounts to about 2.94 m/kwh; allowing 0.3 m/kwh for operating and maintenance costs, 1.76 m/kwh of the 5 m/kwh generating cost represents capital charges. At a 10 percent fixed charge rate and 85 percent capacity factor, this capital charge would imply a total investment cost for a 300 MWe fossil-fueled plant of $39.4 million.

This is only 10–15 percent above 1966 U.S. costs,[∥] which would seem a bit low at first glance, but other factors not present for a nuclear alternative may explain this. First, Israel Electric has had some previous experience in fossil-fueled power plant construction so that dependence on U.S. labor, engineering and management services, etc., should be less than for a nuclear alternative. Second, much of the escalation in U.S. costs reflects substantial increases in the cost of labor rather than materials and equipment; since a fossil plant would allow greater substitution of Israeli labor for U.S. labor, costs may be lower than comparisons with U.S. figures would indicate. Third, for a fossil-fired alternative, it might well be less expensive to purchase major equipment items from non-U.S. sources, offering either lower prices or lower transportation costs to Israel.

Accepting these figures as derived, the power credits at 7 percent and 5 percent fixed charges would be 4.5 m/kwh and 4.1 m/kwh respectively. Thus, at 7 percent, the fossil-fueled plant enjoys a negligible advantage—we may as well call it a standoff—and at 10 percent, the fossil plant enjoys a decisive margin. Only at 5 percent would the nuclear alternative be preferred to the fossil alternative.

---

[*] *1966 Kaiser Report*, p. 129.
[†] *1968 Kaiser Report*, p. IV-4.
[§] *1966 Kaiser Report*, p. 52.
[‡] Jackson and Moreland, *Costs of Large Fossil Fuel Fired Power Plants*, U.S. Atomic Energy Commission, TID-22978, April 1966, p. 12.
[∥] *Ibid.*, p. 22.

Thus, the appropriate power credits corresponding to the least-cost single-purpose alternative to be used in adjusting the Kaiser Report estimates are: 3.7 m/kwh (nuclear) at 5 percent; 4.5 m/kwh (fossil/nuclear) at 7 percent, and 5 m/kwh (fossil) at 10 percent. The corresponding water costs per thousand gallons then become: 33 cents (instead of 22.6 cents in the *1967 Kaiser Report*) at 5 percent; 44 cents (instead of 40.2 cents) at 7 percent; and 66.6 cents at 10 percent.

Thus we see the extent to which the assumption of a 5 m/kwh power credit with other than the 10 percent fixed charge rate overstates the proportion of dual-purpose plant costs attributable to power production and understates those attributable to water production. Note especially that this does *not* involve revenue considerations. Indeed, if all costs are realized as estimated by Kaiser Engineers—a big "if"—and if each kilowatt-hour of power can be sold at 5 mills throughout the plant lifetime, then water could be sold at 23.6 cents per thousand gallons (assuming a 5 percent fixed charge rate) and total revenues from sale of power and water will just equal total plant costs. But this equality of revenues and costs would hold for an infinite set of *other* power and water price-pairings as well, so it is not clear that these particular breakeven power and water prices enjoy any special claim to be the true "production costs" in such a dual-purpose plant.

We could proceed to calculate similar adjustments to the figures in the *1968 Kaiser Report;* however, Kaiser's 1968 estimates of water costs differ from those made in 1967 by less than two cents per thousand gallons, or much less than the uncertainty surrounding any of these calculations.

We should address one further assumption common to all three *Kaiser Reports*. All costs previously given (including my adjusted costs) are based on an assumed lifetime plant capacity factor of 85 percent. Actual experience with presently operating (small) U.S. nuclear plants seems to suggest substantially lower operating factors, as does experience with the largest desalting plant (of the type proposed for the MWD and Israel plants)—probably figures nearer 70 percent than 85 percent.* The reason for assuming very high capacity factors is obvious: given the capital-intensive nature of dual purpose nuclear plants, a high proportion of annual expenses are for fixed charges, and the more units produced, the more the fixed charges can be spread,

---

* See my RM-5227, *op. cit.*, Sec. II, X, and an outstanding paper, addressing not only desalting costs but the demand-price relationship for water under Israeli conditions, soon to be published by Paul Wolfowitz of the University of Chicago.

and the lower the unit cost. Only the justification for the assumption of capacity factors in the 85–90 percent range is lacking.*

While it can be argued that our experience is only representative of "early-years" plant performance,† and that a "mature" plant could sustain higher than average capacity factors, thereby raising the "early-years" average capacity, this abstracts from the problems of "order-of-merit" scheduling discussed earlier, and the as-yet unknown performance records to be attained in later years for either single-purpose nuclear plants or desalting plants. The *1967 Kaiser Report* contains a short table of variations in water costs that would result from changes in certain of their assumptions. The effect of a reduction in desalting plant operating factor from 85 to 80 percent per annum would increase water costs by 2.1 cents per thousand gallons (at 7 percent fixed charges). It appears from this table, however, that this assumes that the power side of the plant would still operate at an 85 percent capacity factor, so that the power revenues would be unchanged. Thus lowering both the power plant and water plant capacity factors to 80 percent would presumably add somewhat more than this 2 cents—perhaps 3–4 cents/thousand gallons.

We might also address the question of the accuracy of Kaiser's capital cost estimates. Here, however, uncertainties have increased over time and appear to dominate the minor adjustments that could be made. Each succeeding Kaiser Report has been smaller and less detailed than its predecessor,§ and has placed increasing reliance on adjustment of costs using various construction and equipment cost indices. The net result is to leave too many holes in the cost assumptions for this kind of analysis to be meaningful.

Moreover, devaluation of the Israeli pound and changes in equipment and labor costs in Israel compound the problems of comparison. The devaluation certainly further jeopardizes this specific project, since the relevant measure is the cost to Israel in agorot per unit of water, which must increase as the cost of U.S. equipment priced in dollars is increased by devaluation of the

---

* Especially when the type of nuclear plant to be used must be shut down for refueling for a *minimum* of 3–4 weeks per year (6–8 percent reduction immediately).

† Performance that appears to be repeating for the first of the large commercial plants. Southern California Edison Co.'s San Onofre plant has been plagued by problems, and Connecticut Yankee's 1968 capacity factor was only 66 percent—see *Nuclear Industry*, September 1968, p. 11.

§ This does not imply less accurate or less exhaustive; it simply makes more difficult the gadfly's task of compensating or adjusting estimates. In particular Kaiser Engineers still seem to be assuming $8.00 uranium ore; see *1967 Kaiser Report*, p. IX-2, item 6.

Israeli pound relative to the dollar. Only the fact that most of us find water costs given in agorot per cubic meter hard to work with obscures the *real* increase in the cost *to Israel* of nuclear desalting.

Of rather more interest is the interpretation to be placed on the desalting costs including escalation, presented for the first time in the *1968 Kaiser Report*. The basis for these escalated estimates assumes escalation of all capital costs and contingencies at a rate of seven percent per year from the original mid-1968 estimate date to mid-1973 which is the mid-point of the construction-expenditure schedule.* This five-year period at seven percent means that all investment costs are increased by a factor of 1.4 over the values in the mid-1968 estimate. Operating, maintenance, and fuel costs are not escalated from those of the 1967 analysis, on the grounds that the 5 m/kwh power credit is a 1967 estimate, and in the hope that "real increase in these costs will be matched by equally higher power credit".†

It is good to see a set of estimates with some provision for escalation, however simplified. One might wonder whether 7 percent is the right figure. It might be high in view of Kaiser's report that "the reactor and turbine prices obtained in mid-1967 were checked to determine possible price increases ... to mid-1968. The manufacturer reported there would be *no change* in the previously quoted estimated prices."§ It might be too low in view of MWD's experience (but a recurrence of that catastrophe in full measure seems unlikely in the Israel case).

One thing that explicit consideration of escalation does is to increase the cost of water, and the rate of 7 percent per annum increases it quite sharply. Were we now to apply our previously-developed single-purpose nuclear power credits to Kaiser's escalated costs, we would find them to be still higher. Indeed, at 5 percent fixed charges, escalated costs per thousand gallons (using Kaiser's 85 percent capacity factor) are increased from 37 cents to 46 cents; at 7 percent, from 60 cents to 64 cents; at 10 percent, however, costs remain at 97 cents. The assumption of 80 percent capacity factors would raise all of these costs by a few cents per thousand gallons.

However, we may question whether Kaiser's simple escalation procedures do not overstate the actual effects of escalation. As noted above, Kaiser Engineers use 1967 operating cost data throughout on the grounds that the 5 m/kwh power credit is also a 1967 value which would probably be adjusted

---

* Kaiser Engineers describe it as the time at which "the weighted average of expenditures occurs". *1968 Kaiser Report*, p. IV-4.

† *Ibid.*

§ *1968 Kaiser Report*, p. IV-3.

upward for changes in operating costs. Now, the power credit should be based on the cost of alternative single-purpose power plant generating costs for plants completed at the same time as the dual-purpose plant (i.e., 1975). However, Kaiser's various descriptions do not make plain whether the power credit is based on a hypothetical oil-fired alternative built in 1967, ordered for 1975, or ordered in 1967 (which would be completed in about 1974, rather than 1975). The escalated cost for 1975 operation would be different for each of these possibilities. In any event, one would expect the construction costs of the alternative single-purpose plant for 1975 completion to undergo some escalation during the construction period, so that that part of the 5 m/kwh power credit reflecting capital costs (rather than annual operating costs) should be increased. Or rather, since their power credit is irrelevant to method D, the calculation of alternative single-purpose nuclear plant costs should include provision for escalation on the *same* basis to the date of start of operations.* While my calculations developed above include an allowance for escalation, it is not a flat-rate 7 percent per annum on all investment costs. Therefore, the final water costs calculated above (46 cents at 5 percent; 64 cents at 7 percent; and 97 cents at 10 percent) should be lowered by a few cents per thousand gallons, to reflect equal escalation of alternative single-purpose plants.†

Parametric estimation of current desalting costs

The two case studies discussed in the preceding chapter dealt with well-defined projects and contained only limited analysis of the effects of changes in assumptions.§ It is quite difficult to evaluate these projects under markedly different ground rules, and the absence of data on alternative fossil-fueled desalting plants prevents a comparison of the economics of various types of plants.

---

* The construction period for a single-purpose plant would be shorter than for a dual-purpose plant, shifting the period of escalation and lowering interest during construction.

† However, this adjustment coupled with reductions of capacity factors to 80 percent should produce nearly offsetting changes, leaving unchanged the last figures discussed above: 46 cents, 64 cents, and 97 cents at 5, 7, and 10 percent respectively. The accuracy of any of these estimates is probably something on the order of several cents per thousand gallons.

§ The *1966* and *1967 Kaiser Reports* contain brief tables of the effects of small changes in operating and cost assumptions, as well as water cost estimates for three fixed charge rates.

An interesting report prepared by the British firm Applied Research and Engineering, Ltd., and published by the IAEA in December 1967 provides some help in performing parametric studies of desalting costs.* The study contains sets of charts designed to permit the intelligent layman to calculate estimates of desalting costs for a wide range of water and power outputs, fixed charge rates, fuel costs, etc., for both fossil fueled and nuclear heat sources. In addition, enough explanatory and backup material is provided to allow some modification of inputs treated as fixed in the charts.† Unfortunately, this volume's usefulness is impaired by the same malady that has afflicted so many of the IAEA reports—its estimates of major plant and equipment items are based on manufacturer's price lists or quotations from the 1965–66 period. Massive cost escalation since those price lists were current renders the straightforward use of the charts subject to substantial understatement of desalting costs.

The basic references for the Applied Research and Engineering, Ltd., (AREL) reactor capital cost estimates are the TVA report§ and an August 1966 (British) General Electric Company atomic power equipment price list.‡ I have discussed in detail in my previously-cited study both the TVA report and the several (U.S.) General Electric Co. atomic power equipment price lists that have been made public.‖ A partial derivation of AREL's reactor costs is given in their Appendix II, in which the quoted G.E. price for a 600 MWe reactor is given as $24.9 million. This precisely agrees with the figure given in G.E.'s August 1966 U.S. price list.# This appears to establish this relevant period during which the estimates were made. However, it should be noted that G.E.'s November 1966 price list increased the base price for this reactor to $29.3 million, an increase of almost 18 percent over the August 1966 price.** While this new list price was subject to some discounting initially, the continuing escalation during 1967 brought price levels back to and ultimately above list prices, where they presently remain.

Similarly, the AREL costs for both the turbine-generator portion of the

---

* *Guide to the Costing of Water from Nuclear Desalination Plants*, Technical Report Series No. 80, International Atomic Energy Agency, Vienna, 1967.

† Also, the charts are reproduced on so small a scale that one cannot read off values to better than about ±2¢, so that spurious precision is eliminated.

§ Reference [8], Technical Report 80, *op. cit.*; the TVA report is entitled "Comparison of Coal-Fired and Nuclear Power Plant for the TVA System", TVA, June 1966.

‡ Reference [7], Technical Report 80, *op. cit.*

‖ RM-5227, *op. cit.*

Ibid. Table 8, and G.E. sources listed hereon.

** *Ibid.*

plant and the desalting plant* reflect prices as of two years ago, and now substantially understate prevailing equipment prices. No equipment price data is given for fossil-fueled heat sources, but it would seem reasonable to assume that they used costs representative of the 1966 time-period.

While we will attempt below to provide some incremental allowances for escalation above the costs used in the IAEA report, we may note at the outset that AREL's outlook for desalted water costs is not much more encouraging than that obtained from our evaluation of the MWD and Israel plants. Even for the largest plant size considered (50 MGD in a single module), the highest steam extraction temperature presently under consideration (270°F), and the lowest fixed charge rate (6 percent†), water costs at the lowest steam cost considered (corresponding to a large power output or to low-cost fossil fuel) are about 25 cents per thousand gallons, using AREL's 1966 cost estimates. Indeed, extrapolation of the curves given to zero-cost steam for the desalting plant yields a floor of just under 20 cents per thousand gallons (unadjusted for escalation to date) at a 6 percent fixed change rate.

Let's look at a few examples of estimated water costs using the IAEA methodology, first without attempting any adjustments. We will use the planning factors for the Israel plant where possible, so that the resulting IAEA estimates can be compared to those of the Kaiser Reports for the adjustments thereto.

For a 100 MGD, 300 MWe nuclear fueled dual-purpose plant, the IAEA methodology gives estimated water costs of 35 cents per thousand gallons at a 6 percent fixed charge rate, and 42 cents per thousand gallons at 9 percent fixed charges.§ The power credits corresponding to these fixed charge rates are, respectively, 3.5 and 4.2 m/kwh, and the desalting plant economy ratio (performance ratio) is about 7.6 (pounds of distilled water per 1000 BTUs delivered to the brine heaters). These calculations are based on method E rather than the revenue credit method of the Kaiser Reports or the adjusted Kaiser figures (using method D) of the last section. Therefore these numbers overstate the water cost and understate the power credit relative to method D; under method D the calculated water cost would be several cents per thousand gallons lower. However, these calculations are also based on

---

* Principal element of escalation here is due to increases in copper prices, affecting the miles of copper-alloy tubing in the evaporaters.

† Which corresponds to 3.4 percent cost-of-money for a 25 year plant lifetime, or 1.8 percent for a 20 year lifetime.

§ These are based on the same turbine exhaust temperature as the *1967 Kaiser Report* (210°F.), interpolating where necessary.

Prospects for Desalted Water Costs 149

AREL's 1966 capital cost estimates, which are substantially below present levels, and assume only a nuclear alternative to nuclear-fueled dual-purpose plants, and only a fossil alternative for fossil plants. Therefore, the single-purpose alternative is *not* in general the least-cost alternative.

Again from the IAEA handbook, a 1000 MGD, 300 MWe fossil-fueled plant,* on the same basis as the above, is estimated to produce water at about 33.5 cents per thousand gallons at 6 percent fixed changes and a fuel cost of 20 cents per million BTUs; at 35 cents per million BTUs water costs rise to about 45 cents per thousand gallons. Power credits are respectively, 3.3 m/kwh and 5.1 m/kwh, and economy ratios in both cases are about 7.5.

Finally, if one is willing to extrapolate and interpolate freely, one can estimate the water cost under the assumption of very low fossil fuel costs. The presence of data in the IAEA report at 50, 35, and 20 cents per million BTUs makes extrapolation to 5 cents per million BTUs seem less risky than other nearly-zero fuel costs. At 5 cents per million BTUs and 6 percent fixed charges, water costs would appear to be about 24 cents per thousand gallons, while the power credit lies off the scale.†

As we are computing a lot of numbers here, it may be helpful to prepare a table for reference as we consider various adjustments and compensations. Table 2 presents first, the estimates directly from the Kaiser Reports; second, the adjustments to those costs resulting from the application of method D; and, third, the estimates of desalting costs developed above from the IAEA report.

How good is the agreement between the modified Kaiser numbers and the IAEA report? The IAEA numbers are still based on method E rather than D,§ and still use understated capital costs; assuming the latter correction would outweigh the former, raising the IAEA estimates by a few cents, the agreement is not bad.

As for the comparison of fossil and nuclear plants, the IAEA report suggests equality of costs at a 6 percent fixed charge rate at fossil fuel prices a bit above 20 cents per million BTUs. The *1966 Kaiser Report*—the only one to consider fossil fueled alternatives to a nuclear plant—concluded that at a 7 percent fixed charge rate, equality occurred at fossil fuel prices of

---

* Utilizing a low-temperature steam cycle, as the ratio of power to water is too low to justify use of a high-temperature cycle.

† It would appear to be below 3 m/kwh, perhaps as low as 2.7–28. m/kwh.

§ And have no provision for evaluating the least-cost alternative—if the dual-purpose heat source is nuclear, the power credit is based on a nuclear alternative, and vice-versa.

Table 2 Water cost estimates for "the Israel plant"—100 MGD, 300 MWe water cost in cents per thousand gallons: power credit in M/kwh.

	Fixed charge rate					
	5%		7%		10%	
Estimation basis	Water cost ¢/1000 gal.	Power credit m/kwh	Water cost	Power credit	Water cost	Power credit
1967 Kaiser Report	23.6	5	40.2	5	66.6	5
1968 Kaiser Report	25	5	42	5	68	5
1968 Kaiser Report, escalated to 1973	37	5	60	5	97	5
1967 Kaiser Report, p.c. adjusted to method D	33	3.7	44	4.5	66.6	5.0
1968 Kaiser Report, method D	34	3.7	46	4.5	68	5.0
1968 Kaiser Report, escalation to 1973, method D	46	3.7	64	4.5	97	5.0

	Fixed charge rate			
	6%		9%	
Estimation basis	Water cost	Power credit	Water cost	Power credit
IAEA Report, nuclear fueled	35	3.5	42	4.2
IAEA Report, fossil fueled at 35 ¢/MBTU	45	5.1	N.A.	N.A.
IAEA Report, fossil fueled at 20 ¢/MBTU	33.5	3.3	N.A.	N.A.
IAEA Report, fossil fueled at 5 ¢/MBTU	24	2.7–2.8(?)	N.A.	N.A.

23 cents per million BTUs—a close agreement, but for a smaller power output.

The nuclear plant economy ratio from the IAEA methodology, 7.6 is also reasonably close to the value of 8.03 selected in the *1967 Kaiser Report*.

Now let's turn to the problem of adjustments to the IAEA numbers. First, we address the question of investment costs. Data in the IAEA report's Appendix II provides capital cost estimates for the nuclear island. The midpoint of the range* given for a 600 MWe heat source, is about $47 million,

---

* The mid-range value was used in the calculations given above.

based, as previously discussed, on an early GE price list and the TVA report. From the data in the IAEA report's figure 6 and section 4.1.2, one can calculate an estimate of investment cost for the turbine-generator portion of a 600 MWe single-purpose plant. That data works out to about $33 million, or a total of about $80 million for a 600 MWe sngle-purpose power plant (about $133/kw capacity). My own report estimated investment costs of about $94 million ($157/kw) for a plant of similar capacity in mid-1967.* Total investment costs for today's large plants in the U.S. are running about $180–$200/kW, or, say, a *minimum* of $110 million. Thus, AREL's capital costs for power facilities in the IAEA report are considerably understated.

As for the desalting plant estimates in the IAEA report, a recent article by two engineers from AREL presents a simplified summary of the IAEA report,† for a 50 MGD unit and 270°F turbine exhaust temperature: $24 million for an economy ratio of 8, $33 million for an economy ratio of 12.5, and $48 million for an economy ratio of 18.§ Since the Israel plant uses an economy ratio of 8, and at 100 MGD is twice the above capacity, one would expect that, if consistency is present, the Israel plant investment in desalting plant would be on the order of $48 million. In fact, however, the Kaiser Report estimates of investment in desalting plant equipment range from $60–70 million (although this is for a lower steam extraction temperature— 240°F), excluding indirect costs, intake and outfall, and interest during construction.

Interpolation of a set of curves in the IAEA report suggests capital costs (for a plant with a performance ratio of 12.5) of about 60–70 cents per daily gallon of capacity at 270°.‡ For a 50 MGD plant, this would imply total investment of about $30–$35 million, which is in good agreement with the $33 million estimate of Starmer and Lowes noted above.‖

We might assume then that the $24 million estimate of Starmer and Lowes for a 50 MGD plant (economy ratio of 12 and extraction temperature of 270°F) can be extrapolated to $50–$60 million for a 100 MGD plant with the same performance ratio but the lower extraction temperature of the Israel plant (240°F.). (While normally one might expect some economies

---

* RM-5527, *op. cit.*, table 9.

† "Nuclear Desalting: Future Trends and Today's Costs", R. Starmer and F. Lowes, *Chemical Engineering*, September 9, 1968, pp. 127–134.

§ *Ibid.;* these figures do not include interest during construction.

‡ Technical Report 80, *op. cit.*, fig. 14.

‖ It also suggests that perhaps interest during construction is not included in the IAEA report's calculations—a subject on which the report is silent.

of scale, the Israel plant actually consists of four parallel 25 MGD modules so that scale economies will be small, and the reduction in turbine exhaust temperature adds to the cost of the plant; therefore it is reasonable to project somewhat more than a doubling of costs in this case.)

A quick summation of Kaiser Report costs for desalting plant and equipment, and an arbitrary apportionment of indirect and overhead costs yields an estimate for the comparable* desalting portion of the Israeli plant of about $85–$90 million. Thus the IAEA Report's estimates appear to understate capital investment in desalting plant and equipment by about $30 million. This, at a 6 percent fixed charge rate, would raise water costs by about 5–6 cents per thousand gallons.

The underestimates for nuclear and turbine-generator plant and equipment mentioned above would appear to have (first-order) upward effects of about 1–2 cents per thousand gallons on water cost (due to higher steam cost) and of about 0.5 m/kwh on the power credit. Thus, the estimates derived above from the IAEA report, 35 cents per thousand gallons for water and 3.5 m/kwh for power at 6 percent fixed charges, would appear after adjustment for escalation, to be increased to about 42 cents per thousand gallons and about 4.0 m/kwh.

Finally, since the adjustment still leaves the power credit based on method E rather than D, we should adjust the power credit to reflect the smaller capacity of the alternative used in method D. Reducing the capacity of the alternative from 600 to 300 MWe might result in an increase in investment costs of perhaps $50 per kilowatt. At the 6 percent fixed charge rate and 85 percent capacity factor, this is about 0.4 m/kwh increased generating cost. This would raise the power credit calculated above under method E from about 4.0 m/kwh to about 4.4 m/kwh, an adjustment that reduces water costs by about 3 cents per thousand gallons. Thus, after this juggling act, the final estimate on a basis closely comparable to the Kaiser Report assumptions, we find the adjusted IAEA figures to be about 39 cents per thousand gallons and 4.4 m/kwh at a 6 percent fixed charge rate (assuming nuclear to dominate fossil for the alternative). This is a bit better fit with the Kaiser Report estimates for 1967 and 1968 at 5 and 7 percent fixed charges than were the unadjusted IAEA figures.

The preceding analysis has dealt largely with minimum-water-cost conditions. A fixed charge rate of 6 percent corresponds at most to a cost-of-money rate of 3.4 percent plus 2.6 percent for a (25-year) sinking-fund amortization,

---

* That is, neglecting interest during construction.

Prospects for Desalted Water Costs

and it is difficult to imagine how substantially lower fixed charge rates could be arranged, much less justified. Let us turn now to the effects on desalting costs of fixed charge rates that more nearly reflect opportunity costs of capital in the private sector.

Private utility fixed charge rates in the U.S. seem generally to fall in the 10-15 percent range, and in general manufacturing may well be above this level. In less developed countries, interest and fixed charge rates (and rates of return at the margin) may also extend upward from 15 percent or so. Since both nuclear power generation and water desalting tend to be capital-intensive undertakings, it is not surprising to find that costs rise sharply for higher fixed charge rates.

For 15 percent fixed charge rates, the IAEA report leads to the following unadjusted water costs and power credits for plants similar to the Israel plant (100 MGD and 300 MWe): for a nuclear-fueled plant, 59 cents per thousand gallons and 6.2 m/kwh, and for a fossil-fueled plant with fuel at 35 cents per million BTUs, 61 cents per thousand gallons and 5.7 m/kwh.

Adjustments to these numbers along the lines of the preceding analysis would lead to still higher estimates,* but, as these are already quite uninteresting water costs from the standpoint of agricultural usage, it would seem pointless to pursue the subject further. One might guess that adjusted water costs at a 15 percent fixed charge rate would be on the order of 65-75 cents per thousand gallons.

It is in some respects fortunate that the water costs calculated under all of these various assumptions are so high: given the range of uncertainty surrounding these simplified and somewhat arbitrary calculations, if the water costs in cents per thousand gallons were in the low twenties or teens, precision would be much more necessary (though still largely unattainable) and the debate would be far more acrimonious. Indeed, Starmer and Lowes sum up prospects as they see them in the following statement:

> It has been stated previously that the target figure for the cost of water is about 15¢/1000 gal. by the 1980s. Figure 9 shows that even with the fast breeder reactor, the present-day cost is very much higher.
>
> There is a scale advantage to be had by increasing the unit size of the MSF [desalting]

---

* Higher in absolute amount that the adjustments previously calculated for 6 percent fixed charges, since much of the adjustment is for understated capital investment to be valued at 15 percent per annum.

plant. However, the curve of water cost against unit size flattens out at about 50 MGD, which reduces the incentive to build units larger than this.*

That this is not a universally accepted summation will be evident when in the next section we turn to a brief discussion of some proposals that assert that large scale nuclear desalting can provide water at costs suitable for agricultural uses.

On large-scale desalting

During the past 18 months, several studies and proposals for large-scale desalting have been undertaken or advanced. They include a joint U.S.-Mexican study under the auspices of the IAEA, a study of "Agro-Industrial Complexes"—plants producing large quantities of both power and water, together with various power-intensive industries and advanced-technology agricultural undertakings—at Oak Ridge National Laboratory, and, most publized, the Strauss-Eisenhower proposal for large desalting plants in the Middle East. As none of these have been reported in detail, our discussion of them will necessarily be tentative and largely non-quantified.

At this writing, the report of the U.S.–Mexico project is not available; only a preliminary statement of this forthcoming report has been offered.† According to that statement, estimated water costs for a 1 billion GPD, 2000 MWe plant range from 16 to 40 cents per thousand gallons, while generating costs range from 1.8 m/kwh for a 4 percent fixed charge rate (!) to 3.1 m/kwh at a 10 percent rate.§ The 4 percent fixed charge rate, if correctly reported, would appear to break the existing world's record (held by the Israel plant) by a full one percent (although it is not clear whether the associated interest rate will lie below the record 1.8 percent also held by the Israel plant). This is a particularly noteworthy rate inasmuch as cost-of-money rates have been rising for many months, and even the U.S. Government has issued new guidelines for evaluation of Federal water projects,

---

* Starmer and Lowes, *op. cit.*, p. 134; their figure 9 shows minimum water costs (at 6 percent fixed charges) of about 27–28 cents per thousand gallons. They go on to suggest that further reductions in water costs will arise from economies of scale in steam production rather than for desalting plants.

† *Nuclear Industry*, January 1969, p. 57.

§ Presumably, the lower end of the range of water costs pertains to the lower fixed charge rate, and the upper end to the higher fixed charge rates.

raising the interest (*not* fixed charge) rate for such projects from about 3⅛ percent to 4⅝ percent for fiscal 1969.*

A summary report of Oak Ridge National Laboratory's study of agro-industrial complexes was published during the summer of 1968.† It is short (30 pages) and lacking in detailed data, and supporting data has not yet been made available publicly.§ Two figures in that report contain most of the information on projected desalting costs. Figure 10 of that report presents curves indicating investment costs (in 1967 dollars) in water facilities including indirect costs but excluding interest during construction for plants from 100–1000 MGD and performance ratios from 7 to 16. Investment costs for present-day desalting plants are estimated therein to range from 35–40 cents per daily gallon of capacity (at PR = 7 to 10) for plants of 100 MGD capacity, to 27–32 cents per daily gallon of capacity (same PRs) for plants of one billion gallons/day (1000 MGD). The scale envisaged for the agro-industrial complexes is also in terms of thousands of MWe and billions of gallons of water.

Their figure 11 presents curves of water cost versus interest rate for plants of one billion gal./day (1000 MGD); these costs range from about 15 cents per thousand gallons at 3 percent, about 20 cents per thousand gallons at 7 percent, to about 40 cents per thousand gallons at 20 percent.‡ At a fixed charge rate of 10 percent (25 year sinking fund plus interest charges), the implied water cost is 22–23 cents per thousand gallons.

As a check on the reliability of these estimates, we can compare Oak Ridge's investment cost estimates for 100 MGD plants to those of the Israel plant from the *1967 Kaiser Report*. Oak Ridge's estimate would be $35–$39 million for a 100 MGD plant bracketing the 8.03 performance ratio of the Israel plant, while Kaiser Engineers' estimate is $67.56 million (both excluding interest during construction). The *Oak Ridge Report* designates these estimates as "near term" or "10 years hence", as compared to 1975 for the Israel plant, and the Oak Ridge estimates presumably reflect U.S. rather than foreign costs. Yet neither a continuation of the present limited

---

* See Federal Register, 33, No. 249, December 24, 1968 [33 FR 19170].

† *Nuclear Energy Centers, Industrial and Agro-Industrial Centers, Summary Report*, Oak Ridge National Laboratory, ORNL-4291, June 1968, hereafter cited as *Oak Ridge Report*.

§ Although the *Oak Ridge (Summary) Report* cites a half-dozen to-be-published reports under preparation.

‡ These calculations assign "all of the cost-saving benefits" of scale to water production, so that they are consistent with the use of method D.

pace of technological progress nor added foreign costs would appear likely to account fully for the considerable difference between these estimates. We might note further that the Oak Ridge report's Figure 9 indicates generating costs of about 3.5 m/kWh for 1000 MWe nuclear plants for 1978 installation at a 90 percent load factor and 10 percent interest rate. At present, for plants for 1974–1975 installation, the comparable estimate would be at least one m/kwh higher,* so the *Oak Ridge Report* must envision some non-trivial reductions in reactor costs as well.

Further comment on the substance of the Oak Ridge report would be both premature and somewhat beyond the scope of this study. There are many interesting aspects to the *Oak Ridge Report* worthy of further study and careful scrutiny. It is to be hoped that supporting data will be made available in the near future so that more detailed analysis can be undertaken and answers provided to the numerous questions raised by a careful reading of the summary report. We shall further address some elements of the ongoing research at Oak Ridge as they impinge on the Strauss-Eisenhower Proposal to be discussed below.

For our purposes, we may note that increasing the scale of desalting capacity by an order of magnitude over present proposals (which themselves are more than an order of magnitude larger than existing installations) appears to lead to near-term prospects of water costs, given low fixed charge rates, no lower than 15–25 cents per thousand gallons, and, quite possibly, several cents per thpusand gallons higher under more conservative investment cost assumptions.† At higher fixed charge rates, water costs from billion gallon per day plants would still be in excess of 30 cents per thousand gallons.

Now let us turn to the Strauss-Eisenhower proposal. This proposal was originally presented by Adm. Lewis L. Strauss, Chairman of the Atomic Energy Commission under President Eisenhower, in a memorandum to Mr. Eisenhower in June 1967. Its contents were made public in articles in several news media,§ and the proposal alternatively became the subject of

---

* Utility executives and AEC officials are presently predicting power costs under 5 m/kwh with somewhat less confidence than they were predicting power costs under 4 m/kwh two years ago.

† In this, we ignore Oak Ridge's estimates for "far term" (20 year) technology, which is inherently speculative at this time.

§ E.g., *New York Times*, editorial pages, July 14, 16, 17, 1967; *U.S. News and World Report*, August 7, 1967.

Prospects for Desalted Water Costs

hearings before the U.S. Senate Committee on Foreign Relations.* It has been further disseminated in articles in popular magazines,† and in newspaper and editorial comment.§

The central features of the Strauss-Eisenhower proposal appear to be summed up by the following excerpts from the Strauss memorandum:

> (a) Let a corporation be formed ... providing that the [U.S.] Government should subscribe to one-half the stock, the balance to be offered for public subscription in the security markets of the world.
>
> b) The amount ... to be initially raised, say $200,000,000, would be used to begin construction of the first of three large nuclear plants...
>
> c) The cost of the plants, beyond the ... common stock, would be financed ... by an international marketing of convertible debentures. They would bear no interest for ... the time to construct the first plant, and could be income debentures depending on market conditions at the time of issue.
>
> d) It [the combined project] should return its cost and then yield income in perpetuity, not only retiring the borrowings incurred but rewarding the governments and individuals with vision enough to subscribe to it initially.
>
> e) The first plant would ... produce daily the equivalent of some 450,000,000 gallons of fresh water.
>
> f) It would also produce an amount of power which, though well in excess of the present needs of the area, would attract industry because of its low cost.
>
> g) Industries such as aluminum and heavy chemicals ... build to take advantage of it [low cost power].
>
> h) It will also be profitable ... to make the nitrogenous fertilizers needed ... A large export trade would result from this item alone.
>
> i) The power would be used [also] to pump fresh water into ... Isreal, Jordon, and other Arab countries...
>
> j) The work of building the great plants, laying the pipelines, constructing holding reservoirs..., powerlines, irrigation ditches, etc., will absorb the unskilled labor of scores of thousands of displaced persons and the skilled work of many hundreds.
>
> k) When the plants are in operation, the greater part of the labor force could be settled in newly irrigated areas...‡

Further information can be drawn from one of Mr. Eisenhower's follow-on articles:

> l) Our proposal suggests three plants ... with a combined production of more than a billion gallons of fresh water per day.

---

* *Construction of Nuclear Desalting Plants in the Middle East*, Hearings Before the Committee on Foreign Relations, U.S. Senate, 90th Cong., 1st Sess., on Senate Resolution 155, October 19, 20, and November 17, 1967.

† E.g., *Reader's Digest*, October 1967, June 1968.

§ E.g., *New York Times*, December 3, 1967, p. 64; *Los Angeles Times*, August 29, 1967.

‡ *Hearings on S. Res.* 155, *op. cit.*, pp. 60–62; explanatory material within brackets added.

m) Since we now know that the cost of desalting water drops sharply and progressively as the size of the installation increases, it is probable that ... water produced by these huge plants would cost not more than 15 cents per 1000 gallons—and possibly considerably less.*

Let us try to evaluate how well of these points hang together. Points (a)–(d) certainly appear to suggest that the majority of the financing is to be obtained through open-market issuance of securities of various types. It is interesting to speculate as to the rate of interest that prospecting investors would require to make funds available for such a venture. A brief perusal of investment pages in any newspaper will indicate that tax-exempt bonds at long term are yielding better than 4.5 percent (for AAA and AA-rated offerings in the U.S.); yields on taxable issues range from 6.25 to over 7 percent; the World Bank is borrowing deutschemarks from German banks at 6.5 percent interest, etc. Since most of the securities to be offered are non-interest-bearing for the first few (5 or 6?) years, their nominal yield would have to be somewhat higher than for normal securities. Also, as it seems unlikely investors would judge the Strauss-Eisenhower proposal to be as risk-free as the undertakings to which the above rates apply, if a substantial risk premium is to be avoided, it would appear that some government† would have to add sufficient guarantees against failure or default. Also, it would appear necessary that all of the Middle East countries to be served by this project regard all phases of the investments as exempt from all taxes—import, property, and income. Unless these conditions are met and the securities are treated as tax-exempt by countries whose citizens choose to invest, there is little possibility of generating the kind of fixed charge rate implied by the water cost figures—one in the 0–5 percent per annum bracket. In the absence of all of these provisions, the fixed charge rate might substantially exceed 15 percent, and several conditions would have to hold to yield even as low as a 10 percent fixed charge rate.

Points b), e), f), l), and m) sum up the rather sketchy specifications of the three dual purpose plants. It seems evident that the maximum desalting plant size is to be 450 MGD. As this is about a factor of 3 increase in capacity over the MWD plant and about a factor of two smaller than the Oak Ridge proposals, costs should fall somewhere between the MND-Israel water cost estimates and the Oak Ridge projections. The latter, let it be recalled, ranged from 15 cents per thousand gallons at a 3 percent interest rate to 40 cents at

---

* "A proposal for Our Time", Dwight D. Eisenhower, *Reader's Digest*, June 1968.

† For projects of the size envisioned, I can think of only one government that might be in a position to offer such guarantees.

Prospects for Desalted Water Costs

a 20 percent interest rate. Moreover, the discrepancy between Oak Ridge projections at 100 MGD and Kaiser Engineer estimates of costs for the Israel plant cast some doubt on the accuracy of this "near-term" projection. Thus, given only the loss of benefits of scale in going back from 1000 MGD to 450 MGD, it would appear that the project interest rate would have to be somewhat below 3 percent to produce water at less than 15 cents per thousand gallons.* Failure to realize the low investment costs Oak Ridge assumes would require a still lower interest rate.

Moreover, a close reading of the Senate hearings suggests that the projection of 15 cents per thousand gallons is based at least in part on the originally reported estimates of water costs from the MWD plant:

> Thus the cost of desalted sea water anticipated at the 150,000,000-gallon [sic] Metropolitan Water District Plant in Los Angeles of about 22 cents/1000 gallons might within ten years fall to around 15 cents/1000 gallons...†

Of course, since the re-evaluation of the economics of the MWD plant in 1968, this kind of citation has been dropped quickly and quietly. The upward escalation of estimated MWD and Israel costs has not led to any increase in projections of either generalized future desalting costs or the projected 15 cents per thousand gallons cost of water from the Strauss-Eisenhower proposal. The nominal future cost of 15 cents per thousand gallons has proved thus far to be much more resistant to escalation than have water costs from any of the more carefully (and repeatedly) evaluated proposals.

Next, we should point out that even if this cost were realized, this represents only the water cost at the plant boundary, not the cost at the point of application. In between are costs for conveyance and distribution systems, for storage systems, and for delivering the water. No estimates of these items are furnished, but reference to other cases suggests these will not be negligible. In the Israel plant case, the sum of $25 million was given as a nominal 1965 estimate for the investment in water conveyance facilities from the plant to the *existing* distribution system.§ Even at the lowest fixed charge rate considered therein, this would add 4 cents per thousand gallons to the reported water cost at the plant boundary. In the MWD plant case, the estimated 1965 investment in water conveyance to an inland blending plant increased the delivered cost of water by 5 cents per thousand gallons.

---

* How this interest rate is to be made compatible with public offering of the securities is left as an exercise for the reader.

† Statement of Alvin Weinberg, Oak Ridge National Laboratory, in *Hearings on S. Res. 155, op. cit.*, p. 78.

§ *1966 Kaiser Report, op. cit.*, Table II and p. 12.

In the Strauss-Eisenhower proposal, as indicated by point (j) a much more extensive system of storage and distribution facilities is envisioned. Not only does this imply substantially higher investment costs, but water storage facilities imply water losses as well (as would open canals and/or leaky pipelines). Any reductions in the produced volume of water at the delivery point due to evaporation, leakage, or overflow would automatically result in higher costs for the water, as production and distribution costs would be spread over fewer delivered units of product. By how much does conveyance and distribution raise the hypothetical 15 cents per thousand gallons? No answer can be inferred from the information given.

Next, the question of power utilization and power revenues comes to mind. First, we should dispose of a widely-circulated but erroneous syllogism that has accounted for much confusion in this area. The syllogism goes roughly along these lines:

Major Premise "Nuclear-fueled dual-purpose plants produce cheap water, witness the MWD plant".

Minor Premise "Nuclear-fueled single-purpose power plants produce cheap electric power, witness the TVA or Oyster Creek plant".

Conclusion "Nuclear-fueled dual-purpose plants therefore produce both cheap power and cheap water".

The error arises in part because power credits used in calculating water costs have been set at levels approximating the price at which the power might be sold or produced under normal conditions, rather than the cost at which it was produced. The excess of power revenue over power production cost then is applied to reduce annual water costs, resulting in lower unit water costs. But if one employs this tradeoff procedure to obtain "cheap" water, one no longer has "cheap" power. Water costs from the MWD plant were based on a power credit of 4 m/kwh, and Kaiser's calculations for the Israel plant used 5 m/kwh; neither of these would be remarkably "cheap" power for their respective areas.

On the utilization of power produced, points f), g), h) and i) are not entirely reassuring. In order that a particular water cost hold, there must exist a market for the power annually generated roughly in line with the power credit assumed.*

To the extent power is used for pumping water, either revenues from other power users must compensate for any price reductions, or else the cost of the pumping power must be borne by water users, thus raising further the

---

* At least, the product of power generated and prices charged must roughly equal the prospective annual generation times the prospective power credit.

delivered cost of water. Since the other points envision pricing power at low enough rates to attract industry, it will be difficult to cover pumping power charges from other power revenues without diminishing the incentive for industrial relocation.

Against the supposed attractive force of low power rates, we should also note some offsetting factors, of which irregularity of power supply is perhaps the largest. Modern power-intensive industrial processes tend largely to require firm sources of electric power supply. The reliability of supply in industrialized countries comes from the presence of an interconnected grid linking many generating facilities and containing sufficient standby generating units to replace immediately any single generating unit that may be forced to shut down.* This security of supply requires a substantially larger installed capacity than the maximum anticipated system demand.

What is the situation with regard to the Middle East proposal? Here each of the three desalting plants would be comprised of (probably) two large reactors, producing some 800–1000 MWe between them (plus excess heat in steam for the desalting plant). Thus each reactor's single-purpose electrical rating would be in the 1000–1200 MWe class (the largest presently offered). The methodology for calculating water costs and power credits makes no allowance for power reserves, so we must look at other possibilities.†

In 1965, the installed capacity in all of Lebanon, Syria, Jordan, Israel, and the UAR was less than 3000 MWe, not all of which is presently interconnected, and reserve capacity was probably small.§ Unless the project envisions construction of extensive transmission interties throughout the area, and a shifting of many of the small, older units in those countries to standby reserve (both of which are *not* costless steps), the necessary reserves cannot be met from present capacity in that region.

Finally, we might consider simply abandoning the customary assuredness of power supply. This would require determining the priorities among power users for suspension of service in the event of loss of capacity, planned‡

---

* Indeed, the criteria are often much stronger than this; spinning reserve on the line equal to the largest two units in service, plus emergency procedures for importing power from surrounding systems in the event of greater need.

† It would appear prohibitively expensive to plan to keep 1000–2000 MWe of the installed capacity idle to form the reserve.

§ The *1965 Kaiser Report* projects only 200 MWe of reserve capacity (and 1400 MWe total capacity) for Israel by 1973.

‡ Each reactor will be shut down annually for a nominal period of three weeks—which may well stretch to six or so if present experience is any guide—for fuel replacement and maintenance.

or unplanned. What would the priorities be? This would, of course, depend on the product mixes and the particular technological processes being used. Presumably, one would not wish to shut down the water plant other than for scheduled maintenance, since the whole point of the project is to produce water, and any reduction in annual water plant load factor inexorably increases water costs.

We are given to understand that the proposed aluminum plant cannot be cut off.* Whether the chemicals plants or the fertilizer plants can be cut off is open to question. If the water plants are operating, can the pumping power be discontinued?

Assuming that some load-shedding scheme could be devised, how do the temporary shutdowns affect production costs in these industries? Given that the sequential refueling shutdowns of the six reactors reduces system capacity by 1000–1200 MWe for at least 18 weeks per year even if the three week-per-unit schedule can be met, how does this affect costs of those industries able to operate only 34 weeks per year?

As for fertilizer production, discussed in point h), we may note the following comments about prospects for ammonia and related nitrate-based fertilizer inputs:

> If a country has natural gas fields, reforming of methane with steam will normally be the cheapest source of the hydrogen required for ammonia manufacture.†

> Table 12 similarly summarizes a number of agro-industrial complexes. Again, the export potential appears poor, ... requiring the most advanced technology to reach an internal rate of return of 10 percent.§

> Taxes, profit, selling expenses, and research and development costs all have to come out of this indicated return.‡

---

* Remarks from the floor by a representative of the Aluminum Company of America, Washington, D.C., November 1968, during the presentation of the Oak Ridge Agro-Industrial Complexes' report, indicate the Alcoa man did not believe aluminum production to be suitable for inclusion in the complex because power supply could not be assured; see *Nuclear Industry*, November 1968.

† *Oak Ridge Report, op. cit.*, p. 13. Electrolytic decomposition of seawater to produce hydrogen for ammonia could be operated during hours of surplus power, thus simplifying *daily* power demand problems (at some cost for underutilization of capital investment in electrolysis). This seems less practical to meet the kinds of seasonal load changes postulated above.

§ *Ibid.*, p. 27. Table 12 is not reproduced herein.

‡ *Ibid.*, p. 20.

Prospects for Desalted Water Costs

In particular, it must be emphasized that the foreign prices ... assigned to industrial products are intended to be representative for countries in which they are needed for internal use; considerably lower prices ... would have to be accepted if one were seeking to export them into world trade.*

Again, some important questions are raised, but the paucity of data both in the Strauss-Eisenhower proposal and the Oak Ridge work does not permit resolution of them. The first quote is particularly interesting for its mention of natural gas, a commodity presently available at zero cost in many of the oil-producing areas of the Middle East.† In many parts of this region, it is presently being burned off at open wellheads, and is viewed very much as a nuisance hindering the extraction of oil. The cost of transporting it to the desalting regions of the Middle East presently under consideration should be investigated, not only from the standpoint of fertilizer production, but to serve as the heat source for power and water production as well, especially in view of the possibilities suggested by the analysis of the preceding section addressing water costs with fuel at 5 cents per million BTUs.

Points j) and k) pertain to the other feature of the Strauss-Eisenhower proposal—employment of refugees. Point j) seems reasonable enough given the ambitious scope of power and water distribution, although one might wonder how "many hundreds" of "skilled workers" there are among the refugees, whose former occupations one would suspect were largely farming and/or shopkeeping.

As to the latter point, a complex proposed by Oak Ridge in a Middle East context, producing 1300 MWe and 610 MGD, seems to project quite small levels of employment:

We have tried to estimate how many workers would be required to man such a complex. Those directly employed on food factory and industrial plant would number 8800. Assuming about four persons (including family) to service each worker, we estimate that the entire enterprise might involve 35,000 people.§

Since the proportions of the plant described therein are about one-half of that envisioned for the entire Strauss-Eisenhower proposal, it would

---

* *Ibid.*

† This is supported by Dr. Alvin Weinberg, Director of Oak Ridge National Laboratory, who remarks in his prepared statement given at the Senate hearings, "Ammonia fertilizer would probably be made from natural gas in the Middle East..." *Hearings on S. Res. 155, op. cit.*, p. 79. The locations of the natural gas fields, however, are remote from the Eastern Mediterranean.

§ Weinberg, in *Hearings on S. Res. 155, op. cit.*, p. 80. He adds, "One would expect that, if the whole idea works, this central activity would attract much subsidiary industry and therefore additional people."

seem that about 70,000 people could be resettled thereby, assuming that one were to rely on the very capital-intensive, very technologically-sophisticated farming methods assumed to hold at all Agro-Industrial Complexes.* So small a number hardly makes a dent in the refugee problem, however, so, if the object is to maximize resettlement of refugees, the technocratic farming operations assumed by Oak Ridge would probably have to be cut back in favor of smaller individual-owner farms to meet the Strauss-Eisenhower objectives. What impact this would have on the efficiency and cost of agricultural production, and to what extent this would reduce the 5–15 percent internal rates of return typical of the Agro-Industrial Complexes considered in the *Oak Ridge Report* are subjects for speculation—or, rather, for more detailed evaluation.

The shift to small farms, presumably lowering efficiency and yields, would raise further questions as to profitability. Both the Strauss-Eisenhower proposal and the *Oak Ridge Report* appear to assume the existence of markets for agricultural production at prices well above current world trade price levels, which assumes the continuation of artificially high price controls in the face of substantial surpluses (the Oak Ridge plant described above is supposed to produce enough food to feed 3 million people).

In support of the technological possibilities, Dr. Weinberg, in the Senate hearings, cites the Gezira cotton project in the Sudan,

> ... started around 1900 when the British decided they had to have a source of cotton in the sterling area and they went into this desert, built a small dam on the Nile, and brought in workers in a very rationalized way.†

This rather direct wording, evoking as it does images of the British Raj and the colonialist policies of the Empire, suggests still another source of difficulty for such technocratic agricultural proposals. It appears to fly in the face of contemporary land-reform desires throughout the less-developed world.

---

* "The yields assumed are those obtained regularly today by the top 20% of farmers in different production areas specializing in the various crops." *Oak Ridge Report, op. cit.*, p. 4. Also, "The general idea ... is to use a highly rationalized agriculture, ... food factories ... at which water and fertilizers are applied at precisely the right time and in exactly the right amounts. If such ... agriculture is indeed feasible ... we estimate that only 200 gallons of water would be needed ... rather than the current requirement of about 2000 gallons per day..." Testimony of Alvin Weinberg, *Hearings on S. Res. 155, op. cit.*, p. 72.

† Weinberg, *Hearings on S. Res. 155, op. cit.*, p. 77.

Many forms of "rationalized" agriculture have existed (and in some cases continue to exist), from the plantation economies of the Old South to the coffee plantations of Kenya, the United fruit plantations in the so-called "banana republics" of Central America, or the rubber plantations of French Indo-China. Although many of the operations in these areas were models of technological and economic efficiency, it seems evident that with the departure of colonial rule this form of production organization is not revealed to have been preferred by the workers (and in many cases, the newly-independent governments).

One wonders whether "food plantations" (to use a less neutral but in many respects more accurate designation than Oak Ridge's "food factory") will turn out to be any more satisfactory if administered by newly independent governments than they were when administered by the former colonial masters. Further, one might cite experience in the Soviet Union with collective farming; productivity and yields of the collective farms tend to be much lower than in the Western world, which proves little,* but *also* much lower than yields from the small privately-owned† farm plots worked essentially as a sideline by collective farm workers. Oak Ridge's yield estimates are extracted largely from large private-enterprise operations, and extension to collectivist operations may not be justified.

Finally, we should further note that a major assumption of the Oak Ridge report bearing on the feasibility of rationalized agriculture is that the use of newly developed strains of wheat and other products together with the application of the best scientific knowledge can reduce water requirements by a factor of ten.§ But this, of course, represents a change in the agricultural production function, and is wholly unrelated to the issue of water desalting. It is in principle as available to present Middle East farmers using existing groundwater and river sources as it would be to a prospective "food plantation". All that the "food plantation" provides is an imposed organizational and managerial function that largely eliminates the need to disseminate this new production technique to individual farmers presently using more inefficient techniques. Whether one wishes to pay the price of a plantation economy rather than paying the cost of disseminating the knowledge to pres

---

* No assertion is implied that these are as rationalized as is suggested in the *Oak Ridge Report*.

† The ownership in a legal sense is unclear, but the major portion of the production is the farmer's to dispose of as he wishes.

§ Weinberg, *Hearings on S. Res. 155, op. cit.*, p. 72, *Oak Ridge Report, op. cit.*

ent farmers—thus in effect *increasing present water supplies tenfold**—is surely an open question.

Beyond this, we should consider what some of the long-run implications of this revolution in agricultual techniques might be. If present agricultural water supplies can over time be made to go ten times farther, the result would seem likely to be a large increase in production of foodstuffs (much beyond that which has already begun to occur in India) and, quite probably, a decline in the arbitrarily high import-level prices used in the Oak Ridge study. Thus, the long-run revenue outlook for the agricultural portion of Agro-Industrial complexes may be rather less favorable than consideration of current prices would suggest.

This concludes our inquiry into the prospects for large-scale desalting in the context of the Middle-East. As is doubtless already evident, estimation of meaningful costs applicable to construction and operation of such ambitious projects is even more difficult than for the smaller dual-purpose plants considered earlier. We have examined some of the central features of these large scale projects, and have explored their internal consistency and external agreement with various off bits of information. The above analysis is certainly not intended to attempt to demonstrate either the feasibility or infeasibility of these proposals; that would clearly be premature. It is intended to raise a number of troublesome economic, technical, and organizational issues that clearly will require much more detailed exposition and analysis before a feasibility judgment can be made.

I should also add that the above analysis is not meant to detract from the value of the Oak Ridge report, or to suggest the feasibility or infeasibility of the complexes they have considered. Indeed, the Oak Ridge report itself neither claims nor denies the feasibility of such undertakings:

Lest one be tempted to draw conclusions too hastily (either pro or con) from such generalized examples, let us briefly mention some of the cautions that should be noted.

As observed earlier, the results are most sensitive to the prices of products and raw materials...

Most of the complexes considered are, especially as for a developing country, relatively enormous from the standpoint of capital requirements, management and operation prob-

---

* That is, Oak Ridge considers only a single effect: if water requirements can be cut by a factor of ten, then, *ceteris paribus*, one can use water ten times as expensive and be no worse off; therefore, desalted water can actually be up to ten times as expensive as conventional sources. However, in the real world, *ceteris paribus* seldom applies, especially in the face of such drastic technological changes. Those presently using ground water at one-tenth the cost of desalted water also need only use one-tenth as much water, and their production costs fall. As this occurs, market adjustments begin and *ceteris paribus* ends.

lems, transportation and other technological support needed, and sociological implementation. The opinion has been expressed, for example, that to get needed information regarding sociological problems will require an order of magnitude larger pilot farm than would be called for by agricultural development problems alone. In fact, the food factory concept would appear to be the reverse of agrarian reform programs in many countries. On the other hand, setting up an operation in a sparsely populated area might be attractive in avoiding complications of existing social organizations and customs. In comparison with the industrialization of existing primitive agricultural systems, there is probably less likelihood of disrupting existing food production.

In an economic feasibility investigation for a specific installation, the proposed industrial or agro-industrial complex must be compared with the best alternative available for doing the same total job. This is beyond the scope of the present study, and conclusions regarding specific economic feasibility should not be extrapolated from it. This is equally true regarding conclusions as to economic nonfeasibility. It is also true as regards conclusions concerning only a portion of the total job.*

Their report concludes with a number of specific recommendations for further studies, of which the most essential for the short run would seem to be:

1) Further more-detailed studies of specific sites, alert to any "special situation" benefits and with careful attention to commodity prices and markets.

2) Cost comparisons with alternate ways of obtaining the same total set of benefits or products.

3) Establishment of an experimental farm for agricultural development in a suitable warm arid region, watered by a small fossil-fueled evaporator, to initiate field experience with cropping patterns in tight year-round relays.

. . .

5) Additional pilot farms for agricultural work and for sociological and economic piloting should follow in due course. It is thought that some of this work would be pertinent to, and regarded in connection with, locales where opportunity exists for watering warm arid arable land from natural rivers in the territory.

. . .

7) Attention should be given to problems of financing and implementation, social and economic effects, transportation and communication and educational and other backups required, distribution and marketing problems, and to regional planning and development.

. . .

9) Construction and operation of a controlled-environment agricultural test chamber to speed up observations on plant behavior in selected soils under selected solar climates and seasons, with control over many variables such as humidity, temperature, wind speed, and CO_2 concentration so that basic phenomena can be more quickly disentangled.†

Thus, about all that can definitely be agreed upon is that Senate Resolution 155 was premature in calling for "the prompt design, construction, and operation of nuclear desalting plants" § in the Middle East.

---

* *Oak Ridge Report, op. cit.*, p. 27.
† *Ibid.*, p. 29.
§ *Ibid.*, p. 1.

To sum up

It appears that, from the clutter of methodologies, assumptions, and data cited and calculated in preceding sections, a few general statements summarizing near-term prospects can be put forth.

First, it appears that nuclear fueled dual-purpose plants of a few hundred MWe power capacity and one hundred million gallons per day or thereabouts water capacity, when evaluated under a methodology assigning all of the benefits of scale to reduce the cost of desalted water, could produce desalted water at as low a cost as 35–40 cents per thousand gallons at the plant boundary, under conditions of base-load utilization and a fixed charge rate not in excess of 5 to 6 percent.

This assumes that actual capital costs for a completed plant will turn out to be reasonably close to the estimations of capital cost, and that escalation of costs during construction will not be substantial. There is, however, little evidence to support either of these assumptions, and inferences both from some proposed desalting projects and from comparisons of after-the-fact costs with initial estimates for many other kinds of projects would appear to support expectations of higher realized costs. Similarly, plant capacity factors used represent nearly the maximum possible rates, and appear to be substantially in excess of levels that have been demonstrated by the (admittedly small) experience to date with single-purpose nuclear plants.

Second, for fixed charge rates typical of privately-financed projects in the United States, the cost of water from the hundred MGD, few hundred MWe dual-purpose plants is roughly doubled, reaching the range of 60–75 cents per thousand gallons at the plant. Such costs appear to lie so far above the cost at which water can profitably be used in agricultural activities that one may question the usefulness of the dual-purpose plant concept.

Third, at the lowest fixed charge rates, fossil fuels begin to look attractive at prices in the 20–25 cents per million BTUs range. At higher fixed charge rates, the breakeven point occurs at even higher fossil fuel costs, but water costs under these conditions remain uninterestingly high. Since parts of the Middle East area are major sources of fossil fuels, including natural gas presently available at well-head at zero-cost, the possibility of importing very low-cost natural gas for both power and water production, and as raw material for fertilizer production, should receive much more attention than heretofore. This appears to offer the only possibility in the near-term for water production costs approaching 20 cents per thousand gallons.

Fourth, these estimates of desalted water costs are for water at the plant boundary. To this must be added explicit and implicit costs for the water conveyance system to the point of use, for storage facilities for water in excess of seasonal needs, for evaporation and loss of water in storage or distribution to the point of final metering, and for the annual operating and maintenance charges on this system, including the electric power to pump water through the distribution system. Since these costs are by and large a function of the specific plant location, existing distribution systems, and location of final users, no "typical" costs can be assigned. Previous studies suggest that even minimal conveyance costs to tie a desalting plant into an existing distribution system increase water costs by several cents per thousand gallons. More complex and extensive systems would tend to raise costs by even greater amounts, further increasing the spread between actual delivered costs, and agriculturally-useful costs

Fifth, expansion of the scale of the plant by something like an order of magnitude above these already large capacities shows some preliminary promise of cutting water costs at the plant boundary by a factor of two, again assuming the realization of estimated investment costs, assumed plant capacity factors, and financial arrangements yielding 5–6 percent fixed charges. The scale of such plants is some two orders of magnitude above any desalting plant yet built, however, and the considerable extrapolation thereby required renders such estimates subject to substantial uncertainty. The large power output of such giant plants requires that power markets be created simultaneously with the plant, introducing many additional planning and organizational problems.

It would appear prudent, therefore, pending a much more detailed and convincing exposition of the economic justification of the one-billion-dollar-plus investment cost of such a very large complex, to view construction and initial operation of one or more of the smaller dual-purpose plants as a requirement for confirmation of the presently speculative economic rationale of these much larger complexes. As experience is gained with plants in the 100 MGD class, one may then consider the merits of substantially larger complexes with realistic cost and operating data to guide the analysis.

I might note in closing that these views have become less controversial since I first addressed the issue of desalting costs some two years ago. Then, the suggestion that desalting costs were not less than 30 cents per thousand gallons was considered heresy, or worse. Now, one finds AEC Commissioner Ramey, certainly no shrinking violet when it comes to forecasting prospects for nuclear energy, remarking

... water costs from large—let us say, 50–100 million gallon per day and 300 MWe—dual purpose, nuclear desalting plants on the order of 30–40 cents per 1000 gallons appear to be achievable in plants built in the near future. While this represents some increase from estimates a year or two ago, it is a figure which should give a green light to properly conceived projects in several selected locations around the world.*

Reference

1. Rand Document D-18456-FF, February 1969. The reprint is the report in its entirety.

* Commissioner James T. Ramey, remarks to a symposium on nuclear desalting in Madrid; quoted in *Nuclear Industry*, Jan. 1969, p. 22.

X

THE BOLSA ISLAND DUAL-PURPOSE NUCLEAR POWER AND DESALTING PROJECT[1a]

Raymond W. Durante

U.S. Department of the Interior, Office of Saline Water, Washington

Introduction

IN MAN'S NEVER ending efforts to insure a sufficient supply of fresh water for current and future needs, the world has witnessed many bold and far-reaching projects. Without question, one of the most daring projects ever to be considered for the purposes of providing new water supply is the Bolsa Island Dual-Purpose Nuclear Power and Desalting Plant. This project was the focal point of attention by potential water-starved areas all over the world. It represented the first attempt by engineers to couple a nuclear power plant with a desalting plant to produce large quantities of electric power and fresh water. From the time that the project was first proposed, it received almost unanimous support from local, state and Federal governments; it was quickly accepted by the public as a worthwhile and desirable adjunct to existing power and water facilities, and it was regarded by most of the technical community as a feasible although challenging undertaking.

Yet at the time of this writing, although the project participants are vigorously pursuing alternative arrangements, it appears that this project will not go ahead on the basis on which it was originally conceived and indeed may not proceed at all. There are of course compelling reasons for this delay and by far the most significant and greatly publicized reason is the increase in costs since the original feasibility studies were made in 1965. While this is true, it must also be obvious that a project such as this which has had unprecedented support by the highest management levels of all the participants and which has been in the planning and negotiation stages for almost three years, is not simply the victim of poor estimating or unforeseen contingencies. Indeed, there are factors which have greatly affected the progress of this project and in turn reflected upon the cost. The purpose of this paper is to examine the cost increase which has occurred between the 1965 estimate and the 1968 estimate, and identify those factors which have contributed to these increases.

Description of the project

Before proceeding with the matter of cost increases, it would be well to briefly describe the project concept and the make-up of its organization and management. Although the participants are presently investigating several alternative arrangements for the plant and organization as a possible means for developing a viable project, the following discussion and in fact this entire paper is based on the overall arrangement as depicted by the negotiated contracts of November 20, 1967. It was on this date that a contract was executed between the United States Government and The Metropolitan Water District of Southern California (MWD) providing the basis of Federal participation and financial assistance of $72.2 million towards the project. Simultaneously, MWD entered into an agreement with the electric power utilities outlining their mutual responsibilities and benefits in the project. We shall now describe the project and the participants as constituted at the time of the signing of these contracts.

The overall project considered the design, development, engineering construction and operation of a large-scale dual-purpose nuclear electric power and desalting plant. This plant included the major facilities generally shown on Figure 1 Simplified Plot Plan, situated on a man-made island (Bolsa Island) to be constructed specifically for this purpose off the coast of Southern California in the area just south of Los Angeles. The proposed design for the island included a peripheral dike of layers of concrete tribar or armor rock which dredged sand to fill the enclosed area, resulting in a useable surface of approximately 35 acres at an elevation of plus 20 feet above mean lower low water level. The nearest point of the island was to be located 2800 feet from the shore, connected to the mainland by a causeway providing an access road and support for the product water pipeline and electric power and control cables. Two nuclear generating units together with all support facilities and equipment required for their operation were to be located on the island. Each unit comprised a light water cooled and moderated nuclear reactor with a thermal rating of approximately 3000 MW(e), reactor containment structure, condensing turbine generators and necessary auxiliary equipment. The two nuclear generating units would be identical or very similar and utilize supporting equipment and facilities jointly wherever possible. A multistage flash evaporator system to be constructed in two phases would be utilized to convert sea water to fresh water. Provisions were made for the necessary inlet and outfall structures sized to handle not only the cooling water for the power plants but also the raw sea water for conversion. The evaporator for Phase I was sized at 50 million gallons per day with additional evaporating units to be added during Phase II to bring the total capacity to 150 million gallons per day. Steam for the desalting plant

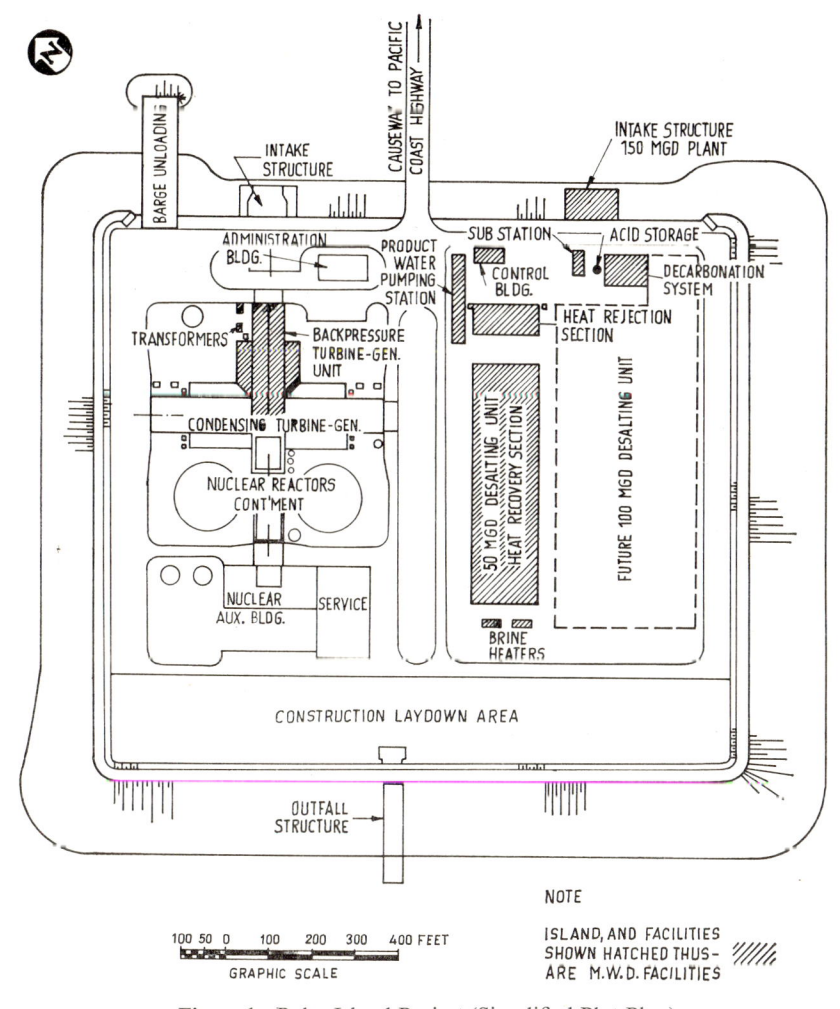

Figure 1 Bolsa Island Project (Simplified Plot Plan).

would be delivered from one or more back-pressure turbine generating units at approximately 260°F and 35 lb per square inch absolute. In addition to the nuclear generating units, back-pressure turbine and desalting plant, there would be equipment and facilities provided which are necessary for their operation and maintenance. These facilities were to be generally used by and available to any of the owning participants, and would include but not be limited to auxiliary buildings and service structures, compressed air system, fresh water supply system, fire protection, heating and ventilating system, waste disposal, handling cranes, general maintenance and repair shops, necessary paving and parking areas, sidewalk, landscaping, lighting, fencing and any other common facilities deemed necessary. High voltage power cables would deliver electric power from the condensing turbine generators and the back-pressure turbine generator to the onshore switchyard for transmission by the utilities to their respective systems. The product water conveyance system consists of a 25 mile 72" diameter epoxy lined steel and concrete pipe extending from the product water sump on the island to the Diemer Filtration Plant, a point of delivery in the Metropolitan Water District distribution system where the water is blended with imported Colorado River water.

Participants in the project

Six diverse organizations agreed to pool their talents and resources and work cooperatively on this project. These organisations and their general responsibilities as depicted by the contracts executed on November 20, 1967, were as follows:

The Metropolitan Water District of Southern California (MWD)

A state-chartered public corporation organized for the purpose of importing and supplying over 1 billion gallons of water per day to 13 cities, 12 municipal water districts and one county water authority in Southern California. MWD was responsible for the desalting portion of the dual-purpose plant. Their participation in this project was for the purpose of establishing a demonstration desalting plant to produce water in significantly large quantities and thereby determine the feasibility of desalting for implementing future water supplies.

The Department of Water and Power of the City of Los Angeles (DWP)

A municipal organization in and for the City of Los Angeles reporting to a board of Water and Power Commissioners. DWP has responsibility for supplying electric power and water within the corporate limits of the City of Los Angeles, and maintains a complete engineering, design and construction organization for this purpose. DWP with a system peak demand of about 2400 MW(e) would assume responsibility for one of the nuclear generating units to be located on Bolsa Island.

Southern California Edison Company (Edison)

An investor owned utility supplying electric power to the Southern California area with a power demand of about 7200 MW(e). Edison, under the project organization for Bolsa Island, assumed the major portion of responsibility for the other nuclear power generating unit.

San Diego Gas and Electric Company (San Diego)

Also an investor owned utility and along with the Edison Company provides electric power to the Southern California area, principally in San Diego County. San Diego was to participate with Edison in the ownership of one nuclear power generating unit, as well as providing additional electric power through the construction of large-scale nuclear plants.

In addition to the generation of electric power, each of the utilities had as objectives for their participation the development of nuclear siting concepts and the demonstration of dual-purpose plant operation.

United States Atomic Energy Commission, Division of Reactor Technology (AEC)

The AEC in the course of its civilian nuclear power development program has directed considerable efforts and interest towards the coupling of nuclear reactors with desalting plants. Through participation in this program, the AEC expected to obtain important knowledge and experience related to not only the coupling of reactors and desalting plants but safety and siting considerations for dual-purpose power and desalting plants and nuclear power plants generally.

United States Department of the Interior, Office of Saline Water (OSW)

OSW, by virtue of its Congressional authorization, is dedicated to the development of information and technology regarding large-scale desalting plants. Its participation in the Bolsa Island project provides the basis for obtaining this technology through experience with an actual operating plant. In return for financial assistance to the project, OSW would receive complete data, technology and operating experience developed through the course of this project.

Neither AEC nor OSW would own any of the facilities constructed or receive any of the products produced. Their participation was based on acquiring technology and operating experience on large-scale dual-purpose plants which they would otherwise have to obtain through other programs.

Contract structure

The participants in the project agreed that two contracts would be established, one between the Government and MWD, and one between MWD and the electric utilities (Edison, San Diego and DWP). Neither document would be subservient to the other and would be negotiated simultaneously with all parties represented during all negotiation sessions. Further, it was agreed that wherever possible identical language would be utilized in both contracts. The contracts contained all of the terms and conditions by which the parties would carry out their responsibilities, their rights and privileges, the basis for operation, management and financing of the project, and the conditions in accordance with which the project could be terminated. It should be noted here that under the termination section was included a clause that permitted unilateral termination by any owner by a given date in the event that the costs to that party should be substantially higher than originally estimated so as to make the project "uneconomical". This condition was accepted as a precautionary measure since firm bids were not available to the potential powers at the time of the contract signing and the only basis on which the project costs could be estimated was the feasibility studies performed in 1965. The intent of this clause was to prevent a situation whereby any party would be compelled to go ahead with the project and thereby incur financial obligations beyond that which they had anticipated. Although there were many other bases for termination of the contract and use of the aforementioned clause was not held highly probable by the participants at the time of

The Bolsa Island Project

the contract drafting, it is singled out here for discussion since it eventually assumed great importance and became the vehicle for eventual termination of the Metropolitan-Utilities Agreement. In addition to the two main contracts, it was planned that other subcontracts would be established for various portions of the work to be carried out, and that these agreements would be between the individual participants and the contractor of their choice.

Project organization

Both contracts provided for the establishment of a Project Management Board. This board consisted of representatives from each of the six participants plus a secretary. The function of this board was to review and control work conducted on the project relating to the total project rather than to the specific responsibilities of any individual participant. This included matters which affected the siting or the design construction, testing or operation of project features involving the coupling or interfacing of water production facilities with power production facilities, or those activities which could cause delay or interfere with the timely progress of the project.

The contracts also called for the establishment of a Project Coordinator. This function was served by the Bechtel Corporation under a contract to

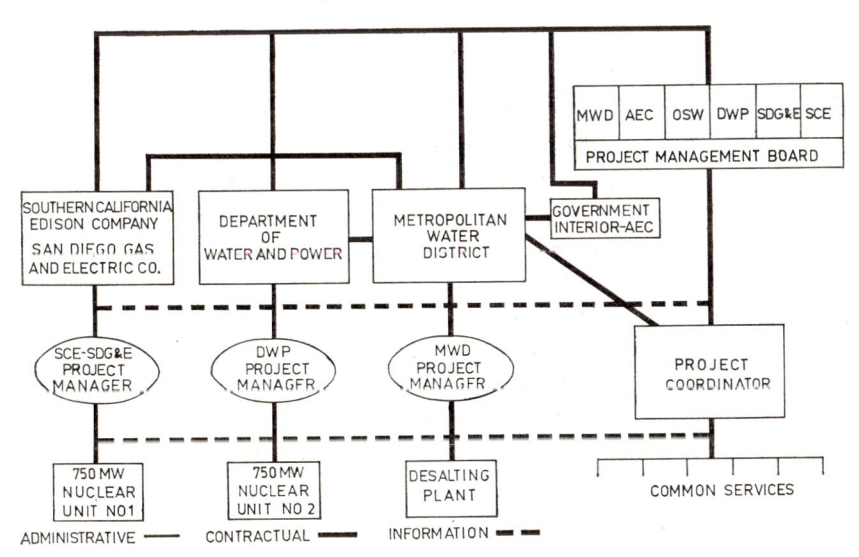

Figure 2 Bolsa Island Nuclear Power and Desalting Plant Project organization

MWD, and their responsibility was to serve as the operating arm for the Project Management Board, particularly in matters related to the overall scheduling and coordination of the total project. It was fully understood by all participants that each of the owners would undoubtedly employ engineering and construction assistance through subcontracts to carry out their specific portion of the project in accordance with their established methods of doing business.

In addition to the Project Management Board and the Project Coordinator, certain other operating groups were established, including the Project Engineers' Committee comprised of the key project engineers for each of the participants; the Project Accounting Committee, and the Public Relations Committee.

Project chronology

In order to assist the reader in understanding the complexity and history of the project, it would seem appropriate to review briefly the steps that have been taken during the past several years leading to the current status. Although MWD, who was the prime mover in this venture, first expressed interest in desalting in 1959 and carried out some contract studies at that time, it was not until April 14, 1964 that their board approved a program to investigate possible desalting plant sites and authorized consultation with appropriate Federal agencies and qualified engineering firms in regard to a possible MWD sponsored sea water conversion project. On August 18, 1964 a contract was signed between MWD and the Department of the Interior and the Atomic Energy Commission for a joint study to determine the engineering and economic feasibility and preliminary design of a combination desalting and power plant in the size range of 50 to 150 million gallons per day and 150 to 750 megawatts of electric power production. The Bechtel Corporation was retained by MWD to perform this study and the work was carried out in three phases: Phase I consisted of a preliminary survey of possible sites; Phase II was a detailed investigation including evaluation of nuclear hazards of not more than three specific sites, and Phase III was a preliminary design and analysis of a dual-purpose power and desalting plant which would be most applicable to the MWD's system requirements. While this study was being made, on April 8, 1965 a joint letter-proposal from the Southern California Edison Company, the San Diego Gas and Electric Company and the Department of Water and Power of the City of Los Angeles was made to MWD. The electric utilities proposed that they participate with MWD in the contemplated plant on a basis whereby MWD would be charged only the incremental cost of a dual-purpose power plant over a single-purpose plant producing an equivalent amount of power. In considering this proposal, costs of water were estimated by the Bechtel Corporation for several different conditions, including single ownership plant producing 150 mgd water and 200 MW(e), and multiple ownership plant in accordance with the electric utilities' proposal producing 1800 MW(e) and 150 mgd of desalted water. On the basis of this comparison, it appeared that a reduction of approximately 25% in the cost of water could be realized and at the same time a considerable reduction could be made in the capital investment required by MWD by joining with the utilities. As a result of this analysis, the MWD Board directed that the balance of the studies consider the utilities' proposal and that a dual-purpose multiple-ownership project be the basis for Bechtel's final report. The Board further directed that consideration be given to the man-made island site near Bolsa Chica State Beach which as a result of preliminary investigations, appeared very attractive not only from the standpoint of access to raw sea water and delivery of product water, but also for the delivery of the power produced. On December 9, 1965, about the same time that the feasibility studies were completed, a report by a task force consisting of representatives of each of the utilities, MWD and the Government was completed. This report developed an overall program of participation by each of the parties if the project were authorized to proceed. At this point it is important to note that while all of the parties were interested and anxious to proceed on a project of this type, no assurance was given that necessary authorizations would be granted by each of the executive bodies, and in the case of the Government, no assurance was available that Congress would appropriate the necessary funds for their participation.

Since a starting date could not be established at that time, capital costs used in the estimates were based upon constant October 1965 dollars with no allowance for escalation. This was done to facilitate the necessary adjustments once the construction schedule had been established. The costs to MWD were also based upon interest during construction computed at a rate of 3.5% per annum, which was slightly higher than MWD bonds were selling for, in that period. Further, costs were based on the premise that engineering construction for the entire project would be performed under a single engineering management type contract.

After numerous discussions between the Secretary of the Interior, the Chairman of the Atomic Energy Commission, the principal executives of the utilities, and the Mayor of the City of Los Angeles and his management staff for the DWP, a Memorandum of Understanding was executed between the Government and MWD on August 19, 1966. This action cleared the way for AEC and Interior's submission to the Congress of the program and request for authorization of Federal funds. Two separate requests for participation, one for the AEC and one for the Department of the Interior, were submitted to Congress, and in October 1966 AEC received prompt authorization and appropriation of their portion of the program. Authorization for Federal participation in this program was reviewed in hearings by the Senate on February 3, 1967 and then by the House on February 27 and 28, 1967 and finally approved in early May. On May 19, 1967 President Johnson signed the bill authorizing participation by the Department of the Interior in the amount of $57.2 million. This added to the already authorized AEC participation of $15 million brought the total Government assistance to $72.2 million.

Meanwhile negotiations were in progress between the parties, and the utilities were involved in the preparation and issuance of bid specifications which would enable procurement of the nuclear steam supply systems. It is significant to note that the controlling factor in establishing the schedule for this project was delivery of the large turbine generators and the nuclear reactors for the power plants. Actual construction of the desalting plant itself would not need to begin until some two and a half years after the project had been started, allowing time for ordering and procurement of the nuclear components and construction of the island. As firm bid prices began to be available, two significant problems arose. First, although the plan was to construct duplicate reactors, each utility prepared its own specifications and conducted separate bid evaluation. Because of certain basic differences in procurement and fiscal policies, the utilities were unable to agree on a single selection of a nuclear steam supply system. Second, the costs to both utilities were ranging considerably higher than originally anticipated. Instead of initiating design and construction activities late in 1967 as planned, the parties extended the date of termination for economic reasons for two successive 90-day periods. During this time, intensive evaluation of costs for the nuclear portion and the desalting portion were conducted, and despite efforts to consider various organizational and plant arrangements which might improve the economic picture, the utilities reached the conclusion that the project as constituted was not economical and they could not proceed. Thus the increased costs had not only delayed this important project, but placed it in jeopardy of complete cancellation.

Costs

As has been stated previously, the basis for the Bolsa Island Project was the feasibility studies performed by the Bechtel Corp. which included cost estimates based upon constant 1965 dollars. As part of the requirement for Government participation in the program, it was necessary to present to the Congress of the United States a complete description of the program proposed, including the costs and technical features which were established in the feasibility studies. It was originally hoped that the project would get under way by mid-1966 and be ready for operation by late 1971. At the time of submission to Congress, however, this schedule was extended a year to accommodate operation in late 1972.

Total cost for the project as submitted at that time was $400.5 million. This consisted of $213.1 million for the nuclear power plants; $153.9 million for the desalting plant, island site and associated facilities, and $33.5 million for a product water conveyance line. The feasibility studies did not consider cost of electric transmission lines; however, during public discussion of this project it was indicated that this cost, based on partial overhead and partial underground transmission, was estimated by the utilities at approximately $43.5 million. While it was clearly stated in all of the formal submissions of this cost that no escalation was included either to bring the cost from 1965 to 1967 or to project them to the actual time of construction, the figure of $444 million became solidly accepted as the "cost" of the project. Using this figure as a basis, Figure 3 shows how each of the major components of this plant increased from 1965 to the current estimates.

Another way of showing the increased costs in the project as related to the individual investment by each owner is shown on Figure 4.

It is important to point out at this time that with regard to the financial assistance from the Federal Government, all Federal funds were to apply to the MWD costs indicated in the table, and are not shown separately since it was determined very early in the project that this assistance would be directed to the desalting portion of this project. The utilities agreed to bear the cost of an equivalent single-purpose nuclear power plant while MWD would assume the cost of a desalting plant plus incremental costs resulting from the dual-purpose aspects of the final plant. It is also important to note that the financial assistance by the Federal Government remains constant in accordance with legislative requirements.

Obviously, these significant changes in the capital cost of the plant resulted

A. ISLAND AND PLANT FACILITIES	Dec. 1965 Estimate	Apr. 1968 Estimate
1. Island and Causeway	23.6	43.6
2. Nuclear Generating Unit # 1	108.8	189.1
3. Nuclear Generating Unit # 2	101.3	182.3
4. Backpressure Turbine Plant	22.3	38.9
5. Desalting Plant	100.9	130.7
6. Land for Switchyard and Right-of-Way for Cable	0.5	2.8
7. Project Coordinator	–	5.5
8. Adjustments	–	7.4
9. Cost of Phasing Construction	9.6	25.8
10. Sub-Total-Phased Plant Construction	367.0	626.1
B. OTHER FACILITIES		
11. Product Water Conveyance System (MWD)	33.5	41.0
12. Power Transmission System Including Land and Right-of-Way (DWP)	33.5	34.0
13. Power Transmission System Including Land and Right-of-Way (SCE)	10.0	42.0
14. Sub-Total—Other Facilities	77.0	117.0
C. SUB-TOTAL—PHASED PLANT CONSTRUCTION	444.0	743.1
D. ADDITIONAL THREE MONTHS ESCALATION TO AUG. 1968		7.4
E. ADDITIONAL OWNERS' CONTINGENCY		15.0
F. TOTAL COST PHASED PLANT CONSTRUCTION		765.5

Figure 3 Bolsa Island Project: Summary of Project Costs by Facility (millions of dollars).

Owner	1965–1966 Estimate	Current Estimate	Difference
SCE/SDG&E	120.3	248.2	127.9
DWP	136.3	239.1	102.8
MWD	187.4	278.2	90.8
TOTAL	444.0	765.5	321.5

Figure 4 Bolsa Island Project: Summary of increase in Project investment by Owner (millions of dollars).

in equal changes in the cost of water produced. Based upon the original estimates for this project it was anticipated that the cost of producing water at the plant site would be approximately 22¢ per thousand gallons. Figure 5 shows the increase resulting in this cost of water due to the rise of capital and operating costs.

It is important to examine the factors contributing to these cost differences and indeed much effort and investigation has been devoted to that task.

Item	Cost of Water ¢ Per 1000 Gallons (150 MGD PLANT)	
1965 Feasibility Report		22
Changes		
Escalation, Anticipated Market Level, Schedule Delays, Design and Scope	+8	
Power Adjustment	+1	
Cost of Money	+3	
SUBTOTAL, UNPHASED DESALTING PLANT	+12	34
Additional Costs due to Phased Construction		+3
TOTAL, PHASED DESALTING PLANT		37

Figure 5 Bolsa Island Project: Change in Costs of Water (At the Plant Site) 1965 to 1974.

Because of the complexities involving the relationship between delay in the project with its accompanying escalation and changes in equipment pricing and plant arrangement, the assignment of dollar values to each factor is not a simple, straightforward arithmetical exercise. For the purposes of this discussion, however, we have shown in the next table a breakdown of those factors which can be singled out and which together bring the total estimate to $765 million (Fig. 6).

Referring to the previous table we shall now discuss each item of cost increase individually:

Escalation of Island and plant facilities

The single most significant factor contributing to the cost increases is the escalation of labor and material prices projected to the time of construction of the project. The original estimate contained in the feasibility studies considered an unphased project of approximately five years' duration, to be completed in 1972 and was based on 1965 prices excluding any allowance for escalation. The effect of phasing the construction of the water plant to allow for four years operating time between the first 50 mgd desalting unit and the second 100 mgd desalting unit; the extended lead time for the delivery of nuclear components, and the schedule stretchout resulted in a project of about 10 years total duration, with the first nuclear plant operating in Sep-

1965 COSTS

Plant, Island & Facilities		357.4
Water Conveyance Lines		33.5
Electric Transmission Lines		43.5
Equipment for Phasing		9.6
TOTAL		444.0

INCREASES TO 1968

A. Escalation of Island & Plant Facilities	152.7	
B. Increase in California Sales Tax	3.0	
C. Higher Power Plant Output	16.0	
D. Market Changes in Nuclear Steam Supply Systems	23.7	
E. Increase in Design Criteria Requirements	17.4	
F. Anticipated State and AEC Licensing Requirements	35.5	
G. Additional Cost Due to Change in Project Responsibility	25.2	
H. Higher Rates for Interest During Construction	16.9	
I. Savings Over the Original Estimate	−(25.3)	
J. Off-Site Facilities:		
Water Conveyance	8.0	
Electrical Transmission Units	33.4	
K. Additional Owners' Contingency	15.0	
TOTAL INCREASE		321.5
1968 ESTIMATED COST		765.5

Figure 6 Bolsa Island Project: Cost Increases (millions of dollars).

tember 1974 and the second phase desalting plant in 1978. Based upon presently available and projected estimates for escalation of labor and materials averaging between 4.5 and 6%, total escalation costs for this project were $152.7 million. It should be noted that the original cost estimate included equipment cost for phasing and not the effects of escalation caused by phasing the construction. The following breakdown shows escalation costs for each of the major portions of the project:

ESCALATION ON ISLAND AND PLANT FACILITIES
(Millions of Dollars)

	Total Island and Plant
Island and Causeway	6.3
Power Plants, Units 1 and 2	
(including Back-pressure Turbine)	82.3
Desalting Plant	49.4
IDC and Other Owners' Costs	14.7
TOTAL	152.7
Product Water Conveyance Facility	8.0

Increase in California Sales Tax

The California State Sales Tax and Use Tax increased from 4% in 1965 to 5% in 1968. This resulted in an increase to the project costs of $3 million.

Higher power plant output

The 1965 cost estimate for a gross plant capacity of 1790 MW(e) was based on standard turbines and nuclear steam supply systems available at that time. In an attempt to standardize available power plant components, manufacturers changed the capacity ratings on available equipment and the utilities decided to procure units of approximately 10% higher electrical output. Although this resulted in a higher capital cost of $16 million, it obviously was advantageous to the utilities, since it added an additional 165 MW(e) and reduced their overall installed cost per kilowatt.

Market changes in nuclear steam supply systems

The 1965 market price level for two reactor systems in the general range considered for this project as compared to the prices received in July 1967 reflected an increase of approximately 40%. In addition, no appreciable discount was offered for the purchase of a second similar reactor as had been anticipated. The net results to the project of this increase is $23.7 million.

Increase in design criteria requirements

After completion of the feasibility studies and as a result of further technical evaluation of the project, it became apparent that additional costs would be required to accommodate changes in design criteria. The major item in this category was the increased island revetment height of approximately ten feet to provide for protection against storm wave overtopping. This was based on results of model tests carried out in 1966. Another significant factor in this category was redundancy in emergency power supply, engineered safeguard systems, pollution and fish kill consideration, and specialized inspection services. The total effect was an increase of $17.4 million.

Anticipated State and AEC licensing requirements

Shortly after the conclusion of the feasibility studies an intensive program of geologic investigation was initiated to determine the acceptability of the site proposed for Bolsa Island. The results of this study indicated that additional costs would be required to provide for differential ground displacement which could result from seismic activity. Also as a result of continuing activities in the area of nuclear licensing, it was now assumed that additional containment augmentation and engineered safeguards would be required for this plant in view of the population density in the proposed area. As part of the detailed investigation performed of the Bolsa site, a special Advisory Committee was formed by the Department of the Interior for geological-seismological conditions pertaining to siting of nuclear desalting plants. Recommendations by this committee led to increased island design criteria to provide increased stability under seismic forces. In general, this category included allowances for additional requirements that might be imposed by the regulatory segment of AEC as a result of completed and detailed studies of such items as tsunamis, wave overtopping, liquefaction of the island fill as a result of seismic activities, the design of the plant for differential ground displacement, containment augmentation and special safety features. The total amount for this category is $35.5 million; over half this amount includes contingencies and allowances due to requirements of State and local agencies, such as the fishkill problem, atmospheric thermal and saline pollution, and other non-nuclear considerations.

Additional cost due to change in project responsibilities

The feasibility study was based on the assumption that a single engineer-manager responsible to the owners would assume full responsibility for the construction of the island and all of the facilities on the island, including both nuclear power plants. It was determined by the owners that each would assume individual responsibility for his portion of the project and that a Project Coordinator would coordinate activities. This change in management philosophy would result in additional costs of $25.2 million. This is based upon the need for duplicate engineering efforts and a loss of construction experience and material savings in the second power unit as well as the reduced use of common facilities. Combined in this category is the action by the owning participants who determined that the two power plants would be operated by two separate operating crews under a single supervisor. This would require duplication of the facilities and operating procedures, as well as additional cost of project coordination for the interface between the two power plants and the back-pressure turbine and the desalting plant.

Higher rates for interest during construction

The original feasibility studies provided an allowance of $35 million for interest during construction. However, the increase in interest rate of approximately 0.75% resulted in a new estimate of an increase of $16.9 million over the 1965 feasibility studies.

Savings over the original estimate

There were two specific items which showed a net saving over the original cost estimate. The most important of these was the saving due to purchase of the large condensing turbine generators which, according to the 1965 cost estimates, were $46.2 million. In 1967 the utilities received firm bids from the English Electric Company which resulted in a net savings of about $20.1 million on the purchase of these turbines. A further item which resulted in a saving to the project was the availability of the copper-nickel material which is used in great quantity for evaporator tubing from the U.S. Treasury at a fixed market price, which when compared to the current price available at the time of the estimate indicated a possible savings of about $5.2 million for the Phase I portion of the plant.

This then represented an increase of $265.1 million, or 72% over the original estimate for the island, plant and facilities.

Off-site facilities

With regard to the facilities required off-site, such as the conveyance of water and the electric transmission lines, there was included in the $444 million an estimated figure of $77.2 million for the combined water conveyance and power transmission facilities. Because the possibility exists that the electric transmission facilities which were scheduled to be built above ground would have to be placed underground, and adjustment to the estimated costs was required, this, in conjunction with escalation of the water conveyance facility resulted in an increase of $41.4 million for the conveyance and transmission facilities.

As a final item on the chart, an overall contingency suggested by the owners of $15 million was added to the original estimate bringing the new total to $765.5 million, or an increase of 73%.

Other factors

It was stated earlier that this project was to have been a focal point for future planners seeking ways to implement water supplies. One cannot accept the cost increases as applicable to dual-purpose plants in general. There were inherent problems *unique to this project* which in the opinion of the writer are important factors which must be considered in any analysis of the costs. These are discussed only briefly as follows:

Project organization

The project was burdened by a complex organization which required a unique management arrangement between private and public power, a State-chartered water distributor and two agencies of the United States Government. Each of these entities has different management and operating philosophies, procurement policies and fiscal practices. While it must be stressed that top level management attention and effort by each of these entities was devoted to this project at all times, there was no single point of management responsibility, but instead a committee management set up to represent all of the parties. The very nature of each of the participants' business structure imposed considerable difficulty in the administration and management of even some of the smallest details and in many instances caused delay in establishing firm decisions.

Nature of the project

The basic ground rules which appear in the contractual agreements between the parties relate to "achieving specific objectives in order to fulfill individual responsibilities while simultaneously coordinating overall efforts to insure the success of the total facility". Such a goal is difficult to achieve in a project which combines the demonstration aspect of large-scale desalting technology, which heretofore had existed only in small plants, with the commercial aspect of electric power production based uppon hard economics and firm availabilities. One of the basic reasons for undertaking this project was to develop information and operating experience that could lead to optimum arrangements for dual-purpose plants, yet answers to questions such as those related to incremental costs and power entitlements were needed beforehand in order to properly negotiate a contract which would initiate the project. Consequently, until more is known regarding dual-purpose operation, future projects may have to proceed on the basis of a more flexible contractual arrangement between the parties.

Siting problems

This project undertook many firsts, especially in the area of nuclear power plant siting. The construction of a man-made island for the express purpose of accommodating nuclear power plants is an entirely new concept and nothing like it has ever been undertaken before. This was further complicated by the fact that the area around the proposed site is in an area of known seismic activity. Previous attempts to locate nuclear plants in such areas have met with difficulty. The proximity of the proposed site with the large population concentration of the Los Angeles area, introduced more stringent siting requirements. All of these factors led the participants to the conclusion that additional nuclear safeguards would be required and more important, considerable delay could occur in acquiring the necessary construction and operating licenses.

Size

While it was generally anticipated that all equipment would be of proven technology, there were some important technical pioneering aspects to the project. The construction of two large reactors side by side coupled through a back-pressure turbine to a desalting plant had never been attempted before. The desalting plant, which was to produce 150 mgd of water was several orders of magnitude larger than any existing plants at the time. The artificial island, even discounting the problems of safety, represented not only an interesting engineering challenge, but also a unique legal problem with regard to ownership and financing. There were many other aspects brought about by the size of this project which although straightforward in appearance represented significant problems and potential delay.

Phased construction

The decision to build the desalting portion of this plant in two phases with an allowance of almost four years between operating dates of the first phase and the second phase introduced many complications as well as aggravated the problems of cost escalation. The problem of operating a plant sized for 150 mgd at one-third this capacity created inconsistencies with the interface between the nuclear generating units and the back-pressure turbines and necessitated additional equipment to accommodate the situation. While there are some advantages to be gained in use of advanced desalting technology by extending the period of time, it would seem that future projects would benefit through the construction of the entire desalting plant at one time.

The above points have been brought out to emphasize that the Bolsa Island Project was a unique and singular situation and that failure to proceed with this project as originally constituted is not to be constructed in any way as an overall indictment against desalting or dual-purpose plants. There is still strong conviction that desalting will eventually play a major role in providing incremental water for areas similar to Southern California. The cost estimates for this project have risen greatly but so too have cost estimates for other alternative means for providing water. In particular, the factor of an escalation which was such a heavy contributor to the cost increase is more obvious as related to construction of aqueducts, dams or other high labor ratio projects, and the normal lead time for desalting project would be considerably less than that required for a large aqueduct. Furthermore, it has been shown that the cost of water will go up and the quality will go down for water imported from greater and greater distances, whereas desalting gives promise for not only a decrease in price due to advances in technology, but it provides the opportunity to put the water supply source where it is most needed and that is near the consumer.

The participants of the Bolsa Island project have reaffirmed their belief that desalting still offers attractive possibilities for solution to the water problems of California and have entered into a Memorandum of Agreement to continue to investigate ways by which alternative means could be found to construct a plant that would demonstrate both the technical and economic feasibility of dual-purpose power and desalting.

Bolsa Island project to be restudied

The Secretary of the Interior Stewart L. Udall and Atomic Energy Commission Chairman Glenn T. Seaborg announced today that alternate plans are being examined for a nuclear power and desalting plant for Southern California in view of the fact that the Bolsa Island project as presently constituted has been determined not to be economic for all participants based on the presently estimated cost of $765 million. Agreement was reached by the participants at a meeting today that the project concept should be continued and efforts were initiated to explore means by which the project could be reconstituted on a more economic basis. At the same time all the participants reiterated their interest in the construction of nuclear power and water desalting plants, including the use of island sites, to meet the growing need of Southern California for these essential resources.

Mayor Sam Yorty of Los Angeles and representatives of the City's Department of Water and Power, Jack K. Horton of Southern California Edison, C. M. Laffoon of San Diego Gas & Electric Company, and officials of the Department of the Interior and AEC, participated in these high level conferences which concluded today. Secretary Udall and Chairman Seaborg also conferred by phone with Joseph Jensen, Chairman of the Metropolitan Water District.

Both Secretary Udall and Chairman Seaborg reaffirmed enthusiastic support for the concepts embodied in the project and the impact they will have on the future of Southern California and the country.

The original plan called for construction on a man-made island of two nuclear power reactors capable of generating 1,800,000 kilowatts of electricity and a desalting plant ultimately capable of producing 150,000,000 gallons of fresh water a day.

References

1a. A Paper Presented at the Symposium on Nuclear Desalination, Madrid, Spain, November 1968. Reprint is the paper in its entirety; also reprinted is the press release of the Atomic Energy Commission regarding postponement of the Bolsa Island project.
1. Engineering and Economic Feasibility Study for a Combination Nuclear Power and Desalting Plant, TID-22330.

XI

NUCLEAR POWER AND WATER DESALTING PLANTS FOR SOUTHWEST UNITED STATES AND NORTHWEST MEXICO[1]

Introduction and approach

In recognition of the present and projected needs for water in the regions of Mexico and the United States bordering the southern portion of the Colorado River, a multilateral agreement was signed on October 7, 1965, whereby the Government of the United States of America, the Government of Mexico, and the International Atomic Energy Agency (IAEA) agreed to make a preliminary assessment of the technical and economic practicability of a dual-purpose nuclear power plant designed to produce fresh water and electricity for this general area. The study was to be directed toward the water and power needs and methods of meeting these needs through 1995.

The agreement was signed for the United States by Stewart Udall, Secretary, Department of the Interior; Commissioner Glenn T. Seaborg, Chairman, Atomic Energy Commission; and Jack H. Vaughn, Assistant Secretary of State for Latin American Affairs; for Mexico by Hugo B. Margain, Ambassador to the United States; and Atomic Energy Commissioner Nabor Carrillo; and for the International Atomic Energy Agency by Director General Sigvard Eklund. A copy of the agreement is shown as appendix I-A. Following formal signing of the agreement, the governments of the United States and Mexico appointed study teams with experts in water resources, water desalting, nuclear power, and electricity generation.

The IAEA appointed Dr. Pierre Balligand, Assistant Director of the Nuclear Studies Center in Grenoble, France to be chairman of the group and Ing. Jorge Spitalnik, Division of Nuclear Power and Reactors, IAEA, to be the scientific secretary.

The study team appointed by the two governments and their areas of expertise are as follows:

U.S. section

Dr. Jack A. Hunter, Director, Office of Saline Water, U.S. Department of the Interior, member for water desalting, head of the United States Section.
Dr. David F. Cope, United States Atomic Energy Commission, member for nuclear power.
Mr. Emil Lindseth, Bureau of Reclamation, Department of Interior, member for water resources.
Mr. Milton Chase, Office of the Assistant Secretary for Water & Power Development, U.S. Department of the Interior, member for electricity generation.

Mexico Section

Dr. Carlos Graef Fernández, Director General Nuclear Center of Mexico, member for nuclear power, head of the Mexico section.
Dr. Alberto Barajas, Mexico Nuclear Energy Commission, member for water desalting.
Ing. José Ramos Magaña, Hydrology Commission of the Valley of Mexico, member for water resources.
Ing. Juan Eibenschutz, Mexico Federal Electricity Commission, member for electricity generation.

Alternate team members for both sections are shown in appendix I-B. In addition to the foregoing, experts and observers were called upon from both countries to assist at working meetings of the study group. Names of those individuals are also shown in appendix I-B.

The study team held five formal meetings between December 1965 and September 1968, and in addition held many small group and expert counterpart meetings. Ground rules were developed and a philosophy with respect to program phasing was established. These are briefly summarized as follows:

1) The requirement for the additional water and power for the region under consideration would be derived separately for: (a) municipal and industrial growth based primarily on a projected increase in population, (b) requirements to sustain present levels of agricultural production and to alleviate existing agricultural water deficiencies, and (c) projections of water required by new land to be brought under cultivation to meet demands for increased food production.

2) The Gulf of California would be used as the source of sea water for the

desalting plant; the plants probably will be located near the northern end of the Gulf. Plant sites, however, need not be adjacent to the coast. Site acceptability, in addition to economic aspects, would be based on meteorological, seismological, ecological, hydrological, and oceanographical factors, as well population density and means for distributing the water and electricity.

3) The economic acceptability of supplying water by means of desalting should be based on the complete program extending through 1995. Plants built later in the program, because of technological improvements, should yield lower water costs than plants built earlier in the program. Thus, the acceptability of desalting as an economic method of meeting future water demands would increase with time. However, to develop the technological base from which water costs can be established and at the same time begin to meet area water needs, it would be necessary to initiate first-plant construction at an early data.

4) It was agreed that, for the first phase of the study, a plant based on the current status of nuclear and desalting technology could be considered. It would have an output of 1 billion gallons per day (bgd) (43.8 cubic meters per second) of water and approximately 2000 megawatts of electricity (MWE), using light-water reactors coupled with multistage flash distillation evaporators. Such a plant would be feasible for operation in about 10 years after the decision to proceed is made.

5) For the later phase of the study, plants were based on improved technology which could be assumed to be available at the time of the commitment to design. For example, plants designed between 1980 and 1990 would employ fast breeder reactors and improved evaporators which would reasonably be expected to be developed by that time. Because of possible inability of disposing of all the baseload power that may be available from future dual-purpose plants, it was also decided that water-only or high water-to-power-ratio plants to meet long-term demands would be considered.

6) Since the fixed charge rates and associated interest rates on borrowed capital may vary both with time and the method of financing, it was decided to do parametric calculations on the water and power costs using the various fixed charge rates of 4, 6, 8, and 10 percent. Equivalent fixed charge rates are used for the civil works which have been adjusted to compensate for the different amortized lifetimes of the civil works structures.

7) Since it is not possible to reasonably project escalation for long periods,

all costs in this preliminary study were based upon 1966–1967 price indexes, and no escalation is included.

8) Since the water demands of both the United States and Mexico were indicated to be higher than their quotas allocated for this study, it was assumed that the water output from the first plant would be divided on a 50–50 basis between the two countries.

9) For this study, the allocation of power to the two countries was based on delivery of an increasing share of power in Mexico until such time as this demand reaches the 50-percent level of net power output from the plant. When this level is reached, future power distribution to the two countries would be based on a 50–50 allocation. It should be noted, however, that the demand for power in Mexico might be limited by transmission costs, since some of the potential users of large blocks of power are located at long distances from the proposed plant sites.

10) The allocation of costs between the products, water and electricity, was based on charging to the water plant the cost of turbine exhaust steam calculated as the fraction of the prime steam cost which represents the additional work that could have been extracted by the turbine if the expansion had been carried all the way to the usual condensing temperature (90°F) for power-only plants. Appropriate charges were made against electricity for the steam condensing function provided by the water plant. The power consumed by the water plant and for pumping was priced at the cost of generation.

11) No consideration was given to the civil works required for the distribution of the product water beyond delivery of this water to a single point in the United States, for its portion of the total, and to one or the other of alternative points in Mexico, for their portion of its total.

12) Certain associated costs were not included, such as right-of-way for water conveyance and power transmission, living facilities for construction workers, premium pay, and unusual construction costs associated with other than ideal site conditions, including special nuclear safety considerations for earthquake-proof construction.

The area where California and Arizona in the United States and Baja California and Sonora in Mexico meet (fig. I-1) is a region where there is a continuing and growing need for water to support increased requirements of minicipal and industrial (M&I) users and meet growing agricultural demands.

The Colorado River is the major stream that passes through this area, but because of the extensive use of this river by both the United States and Mexico, even present requirements cannot be met from this source. In addition, wells which are used to supplement the water supply in the region are seriously lowering the water table. There is little possibility of bringing new agricultural land under cultivation because of the shortage of water in the region. Therefore, it is desirable that early consideration be given to a means of providing sufficient quantities of water to meet the future needs of water users in the four-state region.

As shown in chapter II, estimated "new" water requirements for the region will total approximately 4.5 bgd (197 m^3/sec) by 1995. This total requirement is made up of 100 million gallons per day (mgd) (4.4 m^3/sec) for M&I growth in Mexico, 1.2 bgd (52.5 m^3/sec) to satisfy the ground water overdraft and new agricultural development in Mexico, 1.4 bgd (61.3 m^3/sec) for new agricultural development in the United States, and 1.8 bgd (78.9 m^3/sec) to satisfy groundwater overdraft for authorized agricultural uses and M&I growth in the United States area. It should be noted also that the total rate of increase for both countries is extremely large, yielding a requirement of almost 7 bgd (307 m^3/sec) in the year 2000.

With the development of desalting technology to the point where large plants capable of delivering hundreds of millions of gallons of fresh water per day are considered a reality, it was apparent that this technology offers an approach to meeting future water demands of the region. By the employment of nuclear reactors producing low-cost thermal energy in conjunction with the desalting plant, these water demands can be met at a cost which, though it may not be as low as the prices being paid for natural water, may be competitive in terms of the value received. Also, the use of dual-purpose nuclear desalting plants permits the generation of electric power sufficient to meet a portion of the growing demands for this commodity at a competitive cost.

This study includes an analysis of the water and power areas served; a study of water and power requirements; an analysis of areas which are considered most suitable for plant location; a technical analysis of nuclear and desalting plant concepts to meet water and power demands; an analysis of water conveyance systems; an analysis of power marketing problems; and economic studies concerning total plant capital investment, operating costs, total annual costs, and unit costs of product water and power.

This study consists of two phases. The first is a near-term phase considering those technologies which, through application, would be able to yield a dual-

Nuclear Power and Water Desalting Plants

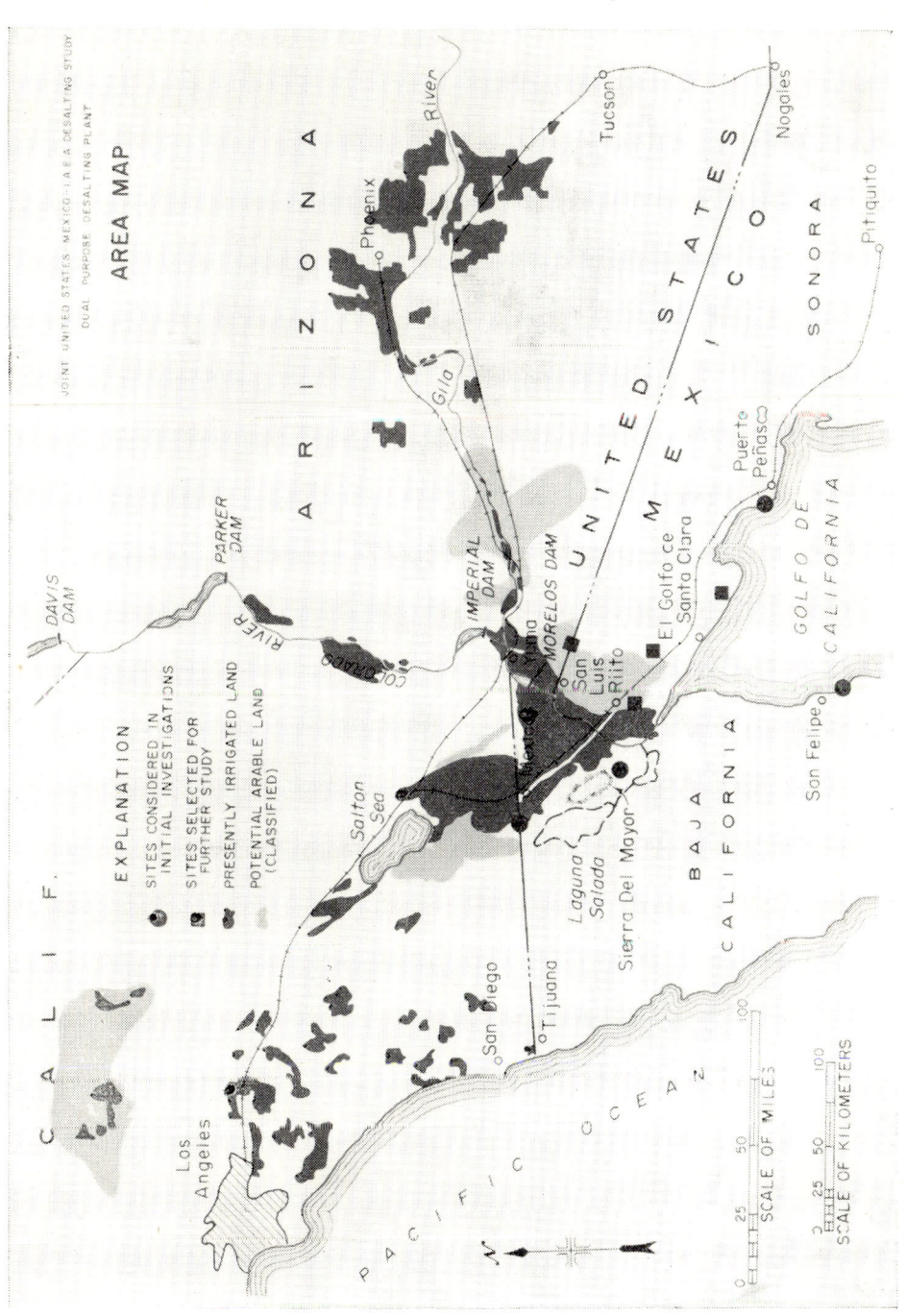

Figure I-1

purpose desalting plant by approximately 1980. The second phase considers plants going onstream in the 1990 time period, when advances in technology are such that improved reactors and desalting systems would be available. As mentioned previously, the initial phase plant system was designed to produce 1 bgd (43.8 m^3/sec) of water together with a power output of approximately 2000 MWE. For the latter phase, variable power-to-water ratios were considered, since it may not be practicable to market all the power which would be produced from the usual dual-purpose plants. Also, for the second phase of the study, plant complexes were considered wherein one site may have more than 1 bgd (43.8 m^3/sec) of water output. Of special interest is the situation of a complex large enough to justify its own onsite nuclear fuel reprocessing and fabrication facility.

The general area studied by the team is presented in figure I-1. Since the Gulf of California is the nearest source of water for a desalting plant, the investigation was directed toward plants located at sites which could employ these waters, either directly or through intake canals from the Gulf. The following sites were considered in the study: El Golfo de Santa Clara, San Luis Rio Colorado, a location near Riito, the locale southeast of Sierra del Mayor, Mexicali, Puerto Peñasco, and San Felipe.

Since the Puerto Peñasco and San Felipe sites are located at long distances from potential users of water, the high cost of delivery far overshadowed any possible advantages of locating at these sites. Conseqently, after preliminary screening, they were dropped from further study. The site near Mexicali also was dropped because of the need for extremely long sea-water intake and brine conveyance systems and unfavorable geological factors.

The desirability of locating a plant south of Sierra del Mayor was complicated by the requirement for either a long intake canal dredged through the existing salt flats or the construction of an intake canal to take sea water from near San Felipe. Either approach was too expensive to be considered in detail.

In view of the foregoing, major study efforts were directed toward sites in the vicinity of Santa Clara and San Luis Rio Colorado. Since water and power costs associated with plants located at these two sites would most likely bracket those for a site south of Riito, detailed power and water cost data were not generated for this case. However, capital and operating costs were developed for the Riito site and are presented in this report.

It should be noted that two site studies were conducted for the El Golfo de Santa Clara locale. One employed sea-water intakes in the area where there was minimum turbidity due to tidal action. The dividing line between high and low turbidity is at about 31°35′ latitude or about 5 miles southeast of El Golfo de Santa Clara. The second site, about 20 miles northwest of the first, employed intakes located in turbid water. The problem that might result from the high solids contents of the turbid water is of unknown magnitude, but this location results in a shorter overall water handling system with consequent lower costs.

The San Luis Rio Colorado site made use of these same two sea-water intake lines, the northern-most, high water-turbidity one following a beachside route and the other low-turbidity intake a highline route. These two join at a point about 35 miles (56.4 km) south of San Luis Rio Colorado, employing a common intake line from there to the water plant. The brine outfall enters the gulf about 10 miles south of Riito.

The product water costs for the San Luis Rio Colorado beachside route and the two El Golfo de Santa Clara routes are essentially the same within the uncertainties of the cost estimates.

The general approach to determining the water requirements for the area under consideration was to project both a minimum growth assuming no forced development, and growth based on accelerated agricultural development. The first of these conditions was based on M&I growth; the agricultural demand remained at the 1967 level, except for those undeveloped agricultural areas with a presently available water supply with the demand being based on maintaining productivity and water quality. The only exception to this condition concerns a possible reduction in agricultural water demand because of improvements in water conservation techniques. The second approach considered new land development predicated on the requirement to meet increasing needs for agricultural products. The amount of land to be so developed is difficult to predict because more effective fertilizers, improved varieties of seed, and more efficient irrigation methods may lead to increased productivity of present land to the extent that major new land development is not required.

Although new water can be supplied to the regions, it is necessary to establish the quantity and quality of water which can be marketed. This region is bounded by the practical limits of water transportation economics and use and is in an area in which water resources have been degraded in quality and depleted in quantity through extensive overuse of surface water and continuing depletion of the underground water supplies for irrigation. Associated with this intensive use is the problem of increasing salinity, increasing the water needed for soil leaching, and possibly forcing a change to lower value crops. Thus, it is generally apparent that the presently available sources of water cannot continue to meet the needs of the region under existing practices. The land contiguous to the present irrigation districts is well suited to agricultural use if the necessary water is made available. This land is quite productive, has the capability of growing high-value crops, and could become a major new food source if developed and placed under cultivation.

In addition, as previously noted, the region of interest borders the Colorado River both in the United States and Mexico. The towns and cities, such as Yuma, Mexicali, and San Luis Rio Colorado, are growing rapidly and demand increasing quantities of water for municipal and industrial purposes. Industry in the area is growing, and although it is not at present heavy industry with its attendant large demands for water, it is expected that in the aggregate there will be uses for water in large quantities. Also, the growth of the overall four-state area will add to the demand for more and higher quality M&I water.

The cummulative effect of these various factors is to impose increasing demand on a limited water supply, tending to further aggravate the water shortage now existing in the region.

The other important factor in determining the amount of marketable water is price. An attempt was made to develop this information by assuming that the price of water to the user would be the cost of its production and delivery, ultimately in the range of 10 to 20 cents per thousand gallons. The difficulty with this approach is that the price which the user pays for natural water often bears little relation to the cost of the delivered water because of various subsidies. This situation applies to both the United States and Mexico. It is also true that water demand is not normally based on a single price, but on a price which the user can afford to pay according to the intended use. For example, high-value crops may support a higher price for water than one is accustomed to paying for water used in growing low-value crops. Also, many M&I uses require water of high quality for which the user is willing to pay a higher price. Thus, quantitative demands for water should be predicted on a range of prices rather than a single fixed price. An important future phase of the program would be to develop more and better information in this regard.

In connection with the pricing of desalted water it was recognized that its higher purity makes it more valuable for either M&I or agricultural applications than the natural water presently available to the region, which is of relatively low quality. In this sense, the range of costs estimated for desalted water, including charges for conveyance, may be quite compatible with the actual costs of new incremental supplies of natural water. Although these prices appear to exceed the amounts now being paid for natural water of lower quality, they do not necessarily exceed the prices which may have to be paid for new water brought into the region. Also, it may be that improved crop yields, growing of high-value crops, or decreased water requirements per unit crop yield (such as by blending) may result from use of desalted water.

The "value" of distilled water in this regard has not been determined, and there is an urgent requirement for additional agricultural experimentation to produce quantitative data on what improvements one can hope to achieve. Such experimentation should cover, in part, the effects of salinity on unit crop yield (both in quality and quantity) for a wide variety of crops, possible decreased water requirements for irrigation, and the variations in time required for crop maturity under the specific soil and climate conditions of the area. From these data, a better comparison can be made of the "value" of distilled water as compared with natural supplies. This is the only realistic way of determining the demand for distilled water as a function of cost. In the absence of good data on the value of distilled water, the chapter VI analysis on the possible benefit in relation to cost of using distilled water for M&I and agricultural uses was based on such information as is available.

Although power demands in the immediate region under consideration are not large, it should be understood that there is a large power-use area within economic transmission distance. Electric power can be supplied to a grid from a nuclear-powdered, dual-purpose plant, but it is necessary that the power be produced at costs such that is marketable in competition with alternate sources.

In the area of power demand, load growth to 1995 was considered, both for the regions studied and for the other United States and Mexican areas which might be involved. It was assumed that a minimum production of power would be 2000 MWE, corresponding to an economically optimum 1 bgd (43.8 m^3/sec), dual-purpose plant, and a maximum of 10,000 MWE corresponding to five such plants. The costs of transmission systems to deliver blocks of power to the southern areas of Mexico was also studied in relation to Mexico's present power requirements, as well as a means for meeting its future growth needs.

The results of the electric power requirements analysis indicate that the full 1700 MWE, or less, of marketable power from the first dual-purpose plant, will find a ready market at production costs resulting from fixed charge rates of dual-purpose plants up to about 6 percent. At higher fixed charge rates, the marketability of this output would be reduced and marginal conditions would result at fixed charge rates of 10 percent or higher.

The general approach in establishing the technical development sequence, overall plant size, types and sizes of reactors and evaporators, pacing of component developments, and scheduling was as follows. The near-term program for the 1980 time period was based on using multiple light-water reactors with a system power level of 10,000 to 12,000 thermal MW and providing the energy to produce 1 bgd (43.8 m^3/sec) of distilled water and approximately 2000 MWE. The reactor technology needed is now available, and the multistage flash technology assumed for the evaporator was based on the technological developments being in hand by 1971.

For the long-term program which extends to 1995, increasing levels of technological advancement were assumed. Reactors considered covered the range from advanced light-water reactors to breeders. Desalting plants were based on combined vertical tube multistage evaporators with advanced components and component improvements. Plant complexes with delivery rates up to 5 bgd (218 m^3/sec) were investigated, and high water-to-power ratio and water-only plants were considered. For the long-term program, the initial onstream date for the plants were based on the time estimated as being required to reach the technological state-of-the-art as called for in the plant concepts plus the time required to construct and start up the plant.

In summary, presently foreseen water and electric power demands for the region appear to justify a plant coming into operation on or before 1980 if civil works are made available to distribute water to the users. A first sucy plant is technically feasible, and successive improvements in technologh should be available as additional plants are required. Each of these improved versions should produce water and power at a lower cost than its predecessor because of advances in the technology, increased construction experience, placing more than one plant to a site, economizing on larger systems, and general improvements and higher efficiencies resulting from increased operating knowledge and experience. Thus, water and power costs from large, nuclear, dual purpose plants should decrease with time in terms of fixed dollar values. Furthermore, as larger and more economic systems are built in the long-range time period, they should have a leveraged effect on average unit costs resulting in an average unit product cost much lower than those attained from earlier plants.

Conclusions and recommendations

The Study Team has concluded that dual-purpose nuclear plants are a technically feasible method of supplying fresh water and electricity to the States of Mexico and the United States that border the lower Colorado River. Such plants could provide the required quantities of water and could enhance the quality of water in the region.

In addition, the study team concluded:

1) An initial plant selected for study purposes, that had water output of 1 bgd and a gross power capacity of approximately 2000 MWE could be constructed by scale-up of present-day technology. After the decision to proceed has been made, between 9 and 10 years would be required to achieve on-stream operation, provided funds are made available on an appropriate schedule.

2) The most promising locations for such a plant appear to be on the border east of San Luis Rio Colorado, near El Golfo de Santa Clara, or between these two locations bear Riito. Final site selection would depend on a more detailed economic evaluation and investigation of all siting factors, such as geology and seismology.

3) The water produced by the first plant is needed by both countries to reduce ground-water overdraft and to aid in supporting projected municipal and industrial growth. The water deficit for the region is projected as increasing from about 1300 mgd in 1980 to 2500 mgd in 1995, assuming no new agricultural development. Under average hydrological conditions, however, presently planned distribution systems in the U.S. would not have unused capacity for distribution of water from this first-phase plant before 1985. In Mexico, additional distribution systems would be required.

4) Based on 1966–1967 price indexes, the total capital cost for the first plant system, including product water conveyance to the major distribution centers, varies from about $850 million to about $1000 million as the fixed charge rate is changed from 4% to 10% when the plant is located at the least-cost site. Different locations could increase the cost at any given fixed charge rate by as much as $250 million. For an advanced system in the 1990 time period, it is estimated that the total capital cost would be reduced 20%. Annual costs for the complete system vary from about $80 million to $180 million per year, primarily depending on plant location and the fixed charge rate.

5) The total water costs at the distribution center from the first plan (1980 time period) would vary from about $15\frac{1}{2}$ cents to 33 cents per thousand gallons for the least-cost plant located as a function of fixed charge

rates. For the greatest cost location, water costs would vary from about 17 cents to 40 cents per thousand gallons. The product water-conveyance contribution to the total costs is about 2 cent per 1000 gallons. In the 1990 time period, technological improvements are projected to decrease water costs from about 6 cents to 10 cents per 1000 gallons.

6) Electricity costs (excluding transmission) for the first plant at 90% load factor vary from 1.8 mills per kilowatt hour at a fixed charge rate of 4% (0.6% interest rate) to 3.1 mills per kilowatt hour at a fixed charge rate of 10% (8.92% interest rate). Breeder reactors, projected for the advanced plants, are expected to reduce electricity costs by about 0.5 mills per kilowatt hour.

7) Owing to the low cost of power produced from these plants and the growth of the power demand in the United States and in Mexico, it appears that the output from the first plant could be marketed. If the ability to market additional baseload power in later time periods were limited, the remaining water demand could be met by water-only systems of advanced design at a relatively small estimated cost penalty.

8) The economic value of desalted water must include consideration of its incremental value in relation to water from alternate sources. The incremental value of this high-quality water can be attributed to such factors as a more dependable warer supply system; the reduced requirement for municipal and industrial treatment; the possibility of increasing crop production per unit of water applies; and the reduced need for leaching. The increment of value cannot be quantified at this time owing to a lack of sufficient data.

9) The use of breeder reactors and advanced desalting processes is expected to result in significant reductions in both power and water costs. The practicability of using this advanced technology depends on successfully conducting development programs, particularly for large desalting systems.

Recommendations

The technical feasibility of large dual-purpose plants for the region has been established. The preliminary economic assessment leads the Group to recommend that further detailed studies be carried out in order to firmly establish the economic practicability. Therefore, to complete the intent of the agreement, it is recommended that the following investigations be conducted by government with the help of private engineering firms where appropriate.

1) The present study is based on 1966–1967 price indexes. In recognition of the impact of changes in the price indexes on the cost of such long-term programs, the probable effect of such changes should be evaluated over the period considered in the study.

2) Verification and refinement of cost estimates should be obtained, including contact with vendors to obtain preliminary bid estimates. Consideration should be given to employing competent engineering firms both in the United States and in Mexico to assist in this effort.

3) In the present study, only preliminary consideration was given to the effect of seismic activity and geological conditions in regard to the location of the reactor and the design of associated conveyance systems. For providing more detailed insight into this important consideration, the three representative plant locations should be studied in greater depth to determine their suitability, taking into account the probable conditions for nuclear licensing, and to adjust the preliminary costs to accommodate the geologic and seismic factors. Three sites recommended for further study are at the border near San Luis Rio Colorado, in the region of El Golfo de Santa Clara, and near Riito.

4) A detailed study should be made of the economics and of the timing for using the water from the first plant, with particular consideration to the earliest possible period when the water could be distributed for effective use. Full consideration should be given to the quality value of the desalted water.

5) The terms under which the power, especially for the first-phase plant, could be marketed should be analyzed and discussed with the utilities of the region.

6) Criteria should be developed, cost estimates should be made, and planning should be initiated for the establishment of an experimental farm program to obtain specific information pertinent to the area regarding the value of improved or high-quality water in the production of food or fiber crops.

7) Each country should consider initiating studies of the overall impact on regional development and economics that would result from the proposed addition of water and electric power into the region.

8) Although milestones in technological development are presented here, it is necessary to define in more depth an overall system research and development program to establish a large component development and testing schedule.

9) An adequate long-range program should be initiated for studying the effect of a desalting plant on the marine ecology of the region. The program should be started early enough to permit the accumulated data to be used in designing the first large plant.

Reference

1. A Preliminary Assessment Conducted by a Joint United States-Mexico-International Atomic Energy Agency Study Team, September 1968. Available as TID-24767 Clearinghouse for Federal Scientific and Technical Information National Bureau of Standards U.S. Department of Commerce, Springfield, Virginia 22151. The reprinted portion is Introduction and Approach, and Conclusions and Recommendations.

XII

PRACTICAL CONSIDERATIONS IN DESALTING AND ENERGY DEVELOPMENT AND UTILIZATION[1a]

James T. Ramey

Commissioner, U.S. Atomic Energy Commission

Introduction

It is indeed an honor to take part in this symposium on nuclear desalting, and to have an opportunity to visit with old friends. I should also like to express my pleasure in being once again in Madrid, one of the world's truly gracious and hospitable cities. We are grateful to my old friend Professor Otero and his colleagues for hosting this conference. We are also pleased that the Honorable Howard Baker, a member of the United States Senate and a strong advocate of large-scale nuclear desalting, is attending this symposium.

Dr. Eklund and his associates in the International Atomic Energy Agency are to be commended for the active support they have given to the advancement of nuclear desalting—by their participation in major studies, by the attention they have given to the growing water and power problems of the various nations, and by their sponsorship of conferences such as this one. These meetings, along with the numerous IAEA technical panels, have contributed significantly to the advancement of effective communication among the many groups which are so vitally concerned with nuclear desalting.

I suspect that many of you are familiar with my long-standing conviction that large scale nuclear desalting is destined to become one of the most important peaceful uses of nuclear energy. On several past occasions, I have had the good fortune to be able to express my views on this subject to international audiences—inluding the Geneva Atoms for Peace Conference in 1964[1], the First International Symposium on Water Desalination, held in Washington in 1965[2], and the International Conference on Water for Peace, held in Washington in May 1967[3]. I recall with pleasure one of the earliest

IAEA conferences on nuclear desalting in September 1963, which I co-chaired with my good friend Jim Carr, then Under Secretary of the U.S. Department of the Interior. I welcome this opportunity to restate the case for nuclear desalting in the light of additional experience and knowledge that has become available to us in the past several years.

To be candid, not all of this experience has been favorable. But it has given us new perspective and a new understanding of some of our problems. None of it, in my view, detracts from the solid basis of our belief that the joining together of the most abundant of the world's energy resources with the unlimited water of the seas to produce fresh water will have a dramatic impact on man's future life on this planet.

Indeed, new studies convince us that we should set our sights even higher. By escaping from the stereotyped idea that new power generation capacity must fit historical patterns of rising power demand, we can considerably enlarge on the capability of nuclear energy to assist the world's developing nations. I am referring to a concept which, by now, is familiar to many of you—the concept of energy centers which combine in a single integrated complex a large nuclear energy source and a group of consuming installations. Such centers would include, for example, electric power generation, desalting, and energy-intensive chemical and metallurgical processes. With these centers, I am hopeful we can break the traditional vicious circle of inadequate demand and high-cost power that has locked the economies of many regions behind what had seemed to be an impenetrable barrier.

The case for nuclear desalting

Because of the doubts which have been raised from time to time, we might well examine once again the fundamental premises of the case for nuclear desalting, and the companion concept of energy centers. Let us try to establish whether this case is still valid under today's conditions. I believe three points are significant:

1) Nuclear energy represents, except for the most favorable fossil fuel circumstances, the lowest cost source of large blocks of energy—and one for which costs are essentially independent of plant location.

Status The validity of this premise is supported by an increasingly impressive volume of solid statistics on the economic choice of energy sources by utilities all over the world. Even recent cost increases for nuclear power, engendered in large measure by the unfavorable market conditions of fully committed manufacturing capacity, have not significantly dampened utility enthusiasm for large nuclear power units. And as market conditions return to normal—and as other temporary reasons of cost increases are corrected—we should again see an improvement in the relative economics of nuclear and conventional power.

2) The economics of desalting are susceptible to major improvement through the same processes of scale-up and evolutionary engineering advances which have reduced the costs of nuclear power and of other modern technologies.

Status Major engineering and development programs in the field of conventional desalting are now reaching the productive stage, and are showing a potential for substantial cost reductions in desalting technology. The completion of the multistage flash evaporation desalting test module in San Diego, California, is serving to demonstrate the technology of large-scale desalting by this near-term approach. At the same time, advanced distillation technology, such as long tube vertical evaporators, also is showing encouraging pilot plant performance.

3) The combination of nuclear power and desalting in large dual-purpose installations serves to maximize the economies inherent in large-scale operations. And large installations let us take advantage of the improved economics of nuclear as against conventional energy sources associated with increased plant size.

Status This factor appears to have growing applicability, as the rapid growth in power consumption and network size in many developing nations has enlarged the areas of potential applicability of large nuclear power plants in the near future.

Taken together, these factors mean that water costs from large—let us say, 50–100 million gallon per day and 300 electrical megawatt—dual-purpose nuclear desalting plants on the order of 30–40 cents per thousand gallons appear to be achievable in plants built in the near future. While this represents some increase from estimates of a year or two ago, it is a figure which should give a green light to properly conceived projects in several selected locations around the world. These are areas where the dwindling availability of conventional water sources, the need for "drought-proof" supplies, and, perhaps, the premium value of high purity water—for use in blending, which

I will discuss later—combine to make the application of desalting especially favorable.

As time goes on, water-short areas will have to reach farther and farther for their supplies, making conventional water sources more and more costly. And technological advances should further reduce nuclear desalting costs. Hence, the long-range economic outlook for large-scale desalting is quite favorable.

When we add to these factors the new dimensions opened up by the energy center concept, I see no basis for pessimism. What I do see is a need to move promptly into the prototype and full-scale plant demonstration phase. We know from our nuclear power experience that the translation of concepts and technology into the design and construction of not just one, but several, full-scale plants is an essential element of the development process.

I believe the time has come to apply this lesson to large-scale desalting. We must cut through the underbrush of economic criteria, opportunity rates of interest, and priorities on capital, and face up to the need of full-scale first-generation demonstration plants.

Against this background, then, let us now examine some of the policy issues and practical considerations involved in building a demonstration plant.

The technological base

General studies

The basis for going into nuclear desalting has been established in a number of general studies on the economic and technical feasibility of large-scale plants in the 1963–65 period. I have in mind, for example, the pioneering report in 1963 on intermediate and large-scale desalting sponsored by the U.S. Office of Science and Technology in cooperation with the Atomic Energy Commission and the Office of Saline Water[4], the study of intermediate size reactors and desalting plants prepared by the AEC in the early phases of the U.S. program[5], and the generalized studies conducted under the auspices of the IAEA during the past few years.

More recently, the energy center concept has opened up a new path for general studies. In this connection, I should mention the challenging, landmark report issued last July by our Oak Ridge National Laboratory. This document is entitled "Nuclear Energy Centers, Industrial and Agro-Industrial Complexes, Summary Report" (ORNL 4291)[6]. The detailed supporting data will be published soon.

Such general studies are important to guide our thinking on the direction which future research and development should take, and to provide a general framework in which more specific case studies can be undertaken. They do not substitute for such case studies, which alone can provide a basis for sound decision-making on individual projects. By the same token, however, they need not—and, in fact, cannot—be undertaken by every country interested in possible nuclear desalting and energy center projects. I would visualize that such general studies will continue to be made, as they have been in the past, by those nations with the most active programs of research and development in these fields. Wherever possible, international organizations, such as the IAEA, as well as representatives of some potential user nations, should participate in these studies to provide the important input of the "users' viewpoint", as well as to insure rapid transmission of the results to the world at large.

In parallel with the general studies, on-going research and development programs are being conducted, both in desalting and in the field of reactors. Definite progress is being made in desalting, and this will be outlined in other papers at the conference. Similarly, progress is being made in the application of reactors to desalting and other industrial processes as will be described later by Mr. Williams, Chief of the AEC's nuclear desalting program. I should note the fact that our Oak Ridge laboratory, because of its background in various disciplines, is playing a principal role in desalting work as well as in reactor development and in integrating the two technologies.

Preliminary preparations for a project

The initial steps in preparing for a project can be critical. Obviously it is of paramount importance to make certain that a firm basis for the project exists. It is vital that close liaison be established as early as possible among the organizations which will participate in the project. This early liaison should achieve a mutual appreciation of the characteristics, goals, and problems of all the cooperating agencies. Not unexpectedly, we have found that next-door neighbors frequently have little appreciation of each others' needs and limitations.

Preliminary study

An essential prerequisite for any proposed plant is a preliminary study, setting forth the conceptual design, preliminary costs, schedules, and other key facets of a project and its relationship to the environment.

After the initial preliminary study has been completed, more detailed engineering investigations can then be undertaken—as was done cooperatively by Israel and the United States—to define more completely the technical and economic aspects. In this fashion, liaison among the participating organizations is developed concurrently with the collection of data.

Because of the complexity of water rights and related laws in many countries, it is necessary to examine the legal structures which could affect the project. Similarly, the charter limitations and financing authority of the participating groups should be fully understood by all parties, and they should be reviewed periodically, as planning proceeds, to prevent overextension of any participant's capabilities.

Evaluation all the potentials

In any preliminary study it is useful always to consider alternate resources. It is important in such a study that the participants realize the unique characteristics of desalting as a new water source, and that its integration with other water resources be evaluated.

For example, I previously have commented on the potential of desalting for "drought-proofing" an area. Investigations reinforce our confidence that desalting can be effective in accomplishing this objective. The addition of desalting plants to handle seasonal "peaking" demands and to increase reservoir storage during drought appears to be an economically attractive way to increase the firm capacity of a water system.

We should remember, too, that it is possible to plan and construct desalting plants in any desired capacity, and on faster schedules than can normally be achieved with large-scale water importation systems. The use of desalting plants as a supplementary water resource, in combination with water importation, provides a totally new flexibility in the planning of water resource development.

Combining the production of desalted water with that of electric power requires that plans for the development of both resources be involved in the project. Various approaches can be taken in unified resource development. These include consideration of peak versus base load power requirements;

the use of hydro-power to supplement power from the dual-purpose plant in peak periods; and the ability to store the desalted water in existing facilities. These factors, along with others, also introduce the need to carefully assess the values to be assigned to the two products of a dual-purpose plant.

If one is considering a large-scale installation, it is essential that the facility fit into regional as well as local plans. In the event a regional plan for water and power development does not exist, the project may act as the catalyst to initiate such planning, and perhaps become the focal point of a regional plan.

A large-scale power and desalting project often can offer significant benefits in such areas as recreation and economic growth to the surrounding community. These other benefits should be identified and evaluated.

One other important factor will be the matter of siting. In some cases the nuclear characteristics may well impose certain restraints. Accordingly, it is advisable to reach agreement on an acceptable site at an early stage in planning.

Determining project viability

Now let me emphasize what may seem like an obvious point, but it is most important: As plans begin to take form, it must be firmly established that the project is, in fact, viable.

Economics

To be viable, the facility must have over-all benefits which are accepted by a broad segment of the community as sufficient to justify the cost. Economics is important, but not the exclusive factor in this determination. Certainly any evaluation should take into account all considerations, including the costs of alternative sources of water and energy, and an expanded resource base. I would like to elaborate on these points, because they are of critical importance.

The economics of alternative power and energy resources should be evaluated objectively on a consistent set of standards. Some critics today compare the cost of desalting with the cost of water from systems which have been in existence for 20–50 years or more. Frequently comparisons are made with subsidized prices for water. And increasing the capacity of these natural water resources often is impossible at any cost. Desalting offers a valid alternative for augmenting available water resources; it should not be considered as a substitute for existing water supplies.

While I'm on the subject of economics, I would be remiss if I did not point out a most significant contribution of large-scale desalting. The costs of "conventional" water resource development, including importation, have constantly increased. Desalting costs, on the other hand, have shown a significant downward trend.

A vital consideration in the assessment of a project is the determination of the value of the water to be produced, including particularly any special values resulting from the high purity of desalted water. Traditional concepts of water value, on careful examination, often are found to be based more on arbitrary prices, not closely related either to the economic value of the water in terms of its contribution to the production of other commodities or to the true cost of producing the water itself. More and more studies show that the value of water in specialized agricultural uses can be far higher than the arbitrary prices and costs long associated with water for irrigation. I am not arguing against the subsidization of water costs—that is a decision to be made by each country for itself and there may be valid reasons for doing so. I am saying that in the evaluation of projects, it is essential that true figures be used for both water value and water cost.

As I have said, these figures should not overlook the potentially large contribution which can be associated with the high purity of desalted water. By blending this water with available water of high salt content, a gallon of desalted water may well add more than a gallon of water to the total supplies available to a country or a region.

Financing

Financing of the project can be critical to its viability. Several forms of financing could well be involved, including equity or debt financing by private corporations, debt financing by public bond issues, and income from the general taxing authority of the participants. One or more government agencies could participate, consistent with the objectives of the project, and the time-flow of this portion of the financing must be planned. Special loans from international lending institutions could well be required.

It is obvious, then, that a comprehensive financial plan must be prepared. It is equally obvious that the first demonstration nuclear desalting plants may well require various forms of financial assistance.

The problems of project financing are magnified when one considers the energy center concept. Continued studies of this concept, which are under way at the AEC's Oak Ridge National Laboratory, will include investigations of the financing problem. This effort may provide extensive information that also would be applicable to the financing of large power and desalting projects.

Organizing the project

As one proceeds to the actual organization of a project, additional factors of critical importance will be encountered. A first precautionary step is to recognize all the parties who will be affected by the installation and which will participate directly.

In the United States, as well as in many other parts of the world, power and water are normally the responsibilities of different organizations. For this reason, the concept of dual-purpose plants—which has as its principal objective the provision of low-cost energy to the desalting plant—will usually bring together two or more organizations with different interests.

And the organization problem becomes more complex when a water project spans national boundaries. Some situations may call for novel arrangements such as the creation of corporations with international ownership—both public and private—functioning under international agreements or conventions. In such cases, we can draw on experience gained in other fields in which international cooperation has been achieved.

A large power and desalting project probably will be a major effort for the participating organizations and will confront each of them with unusual, if not totally new, dimensions of management and administration. A project of this nature cannot proceed successfully if the organizations insist on doing business "in the same old way". It will demand as much imagination and creativity in management and administration as in the technical disciplines.

Unlike technical problems, the problems of establishing effective management of a large-scale, high-cost project cannot be resolved in the laboratory or even with studies. Management will, of necessity, be a real pathfinding effort. It will require tough-minded people who can assume big responsibilities and who have imagination. They must be willing to try new approaches, and to act with boldness. The unique organizational relationships of the project may occasionally require that a few heads be knocked together, but this is true in many undertakings.

When agreement is reached on a technical description of the project, the responsibilities of each participant should then be defined. If there is to be multiple ownership and a division of design and construction activities, then this delineation of responsibilities is most essential. It must clearly set forth the interfaces of the project—interfaces not only with regard to the capital equipment, but also those in engineering and design, construction, and the operation of the plant. The assignment of responsibilities should evolve into written agreements as early as possible.

A single over-all policy organization must be established which properly represents all parties and which can provide general policy guidance. Of even more importance, a single organization—preferably an experienced utility—should be delegated the responsibility for over-all management of the project and this lead organization must have sufficient authority to provide coordinated direction for the design, construction and test operation of the project, and particularly the interplay between the power and desalting aspects. Otherwise the project will bog down under the burden of too many bosses. Special plans which will eliminate or expedite the normal administrative reviews within the participating organizations will be required if the project is to proceed expeditiously.

Most importantly, none of these steps will be truly effective without a willingness on the part of the participating organizations to moderate the management philosophies and preferences they normally would apply to a job which they would undertake unilaterally.

Establishing a schedule for the project

The planning effort could well be a total loss unless a *realistic* and *firm* schedule for the project is established early in the planning phase. Scheduling, as many of you know, can be a delicate matter. If you schedule too far in the future, interest and enthusiasm will lag. On the other hand, an overly optimistic schedule can be the basis for subsequent setbacks which can be fatal to the entire undertaking. A proper phasing of the project with other facilities producing water and power must be accomplished, and the other interests of the project participants also must be satisfied.

Cost escalation

In considering a firm schedule, let's take note of cost escalation. Although this phenomenon has plagued us for many years, it has been receiving increased attention recently throughout the world as economic systems have become more complex and money values have fluctuated more rapidly.

So cost escalation is a problem in any major undertaking these days, and is not just characteristic of large nuclear projects. As costs generally rise, including costs for alternate water sources, we believe the competitive position of nuclear desalting will be enhanced since it will continue to benefit from advances in technology.

Cost escalation was a major factor in the increased estimates for the Bolsa Island project, and these increased costs posed serious problems. As many of you know, this large-scale dual-purpose nuclear plant was proposed for Southern California as a cooperative effort of The Metropolitan Water District of Southern California, the electric utilities serving that area, and the U.S. Government. About half of the increase in the estimated cost was due to cost escalation. Other factors that contributed to the increase were: (1) the proposed island site, (2) its location near a large population center and in a region of seismic activity; (3) a change in the organization of the project; (4) changing from overhead to underground power transmission systems; and (5) increases in taxes and financing costs.

These cost increases made the project, as it had been initially constituted, uneconomic for some of the participating utilities, and we have been studying alternative sites and arrangements. An analysis of the cost increases of the Bolsa Island project will be reviewed in a paper to be presented by Mr. Ray Durante of the U.S. Office of Saline Water at this conference.

International cooperation

From the outset of the United States' program on large-scale nuclear desalting, a major objective has been cooperation with other nations directed toward widespread application of the benefits of this new technology in the water-short areas of the world. We have shared and will continue to share our knowledge and experience with others. We have gone beyond this by cooperating directly with a number of nations in surveys and studies of the applicability of nuclear desalting to specific needs or regions. In carrying out these activities, we have worked closely with the IAEA, whose participation has been invaluable both to insure the widest possible exchange of information in the field as well as to bring to bear the special knowledge and judgment of an international organization. The U.S.–Mexican nuclear-desalting study under IAEA sponsorship and the able Chairmanship of Dr. Balligand and Secretary Spitalnik is a good example of IAEA leadership.

The IAEA has, in addition, as this symposium demonstrates so well, itself taken a broad initiative to disseminate knowledge and increase international understanding of nuclear desalting technology and the plans and programs of specific nations.

Energy center concept

As we seek to encourage the new concept of energy centers, the need for international cooperation and the role of international organization will become even more significant. Although I do not by any means rule out its application in the United States and other highly industrialized nations, the energy center concept is essentially tailored to the special problems and opportunities of developing nations—and groups of nations—particularly in the arid regions of the world. The complexity and diversity of the factors which must be considered in studying and, ultimately, planning energy centers—factors such as water and power demands, agricultural production and marketing, industrial capabilities and the market for manufactured products, financial and even sociological considerations—absolutely require that the broadest possible range of talents and judgments be brought to bear on each potential application.

In these circumstances, the capability of the IAEA to draw on its own and other international sources of knowledge and people will prove invaluable.

In our concept of "energy centers", we have distinguished between the agro-industrial complex, whose emphasis is on increased food production through desalting and the manufacture of agricultural chemicals, and industrial complexes, where the emphasis is on the use of nuclear power for energy-intensive chemical or metallurgical processes. Obviously, intermediate cases between these boundaries can also be postulated.

At Oak Ridge National Laboratory, we are examining the prospects for a very large agro-industrial energy center in the Near East. Such a center would be regional in nature, possibly affecting several national boundaries. Obviously this idea would have far-reaching economic and social implications.

On a more reduced scale, we are looking into the potential for a more limited energy center, or a "partial" energy center, in Puerto Rico[7]. This concept might well embrace power production and some desalting, combined with industrial and perhaps some agricultural enterprises.

It is especially in the field of agro-industrial complexes that I find the most exciting prospects for applications that hold the promise of having a fundamental impact on the economy and development of entire regions. Through the availability of large volumes of water and low-cost fertilizer from an energy center, lands that have lain fallow since the dawn of history may respond to a new kind of intensive agriculture which feeds millions of hungry mouths. Ultimately, perhaps only in this way can we avoid or significantly delay the Malthusian threat of inadequate food for a world population growing at an unprecedented rate.

At first, however, applications of the energy center concept will necessarily be few in number and must be selected with the utmost care. I do not want to minimize the major obstacles of technology, planning, and funding that must be overcome before ground is broken on even one large energy center in the world.

The planning cycle—the time required from conception to completion for projects of this magnitude—is a lengthy one. This makes it all the more important that here, as in the somewhat simpler case of nuclear desalting projects, we must begin now the preliminary steps which will lead us to successful applications in time to meet the needs and opportunities which we will face in the 1970s and beyond.

I look upon this as one of the most exciting challenges of the international nuclear community.

Conclusion

Today we find ourselves at a key point in time. If large-scale nuclear desalting is to reach its full potential, it must proceed with confidence into the demonstration phase. Not one, but several, demonstration projects—using different technologies and in diverse locations—should be started without further delay. The experience gained in these ventures will provide a foundation for further advances. Strong leadership will be required to fully realize the needed demonstrations in view of the broad scope of such projects and the heavy financial commitments involved.

Drawing on the technology and experience already available, the "success story" of nuclear power can be repeated with large scale desalting. Our natural water supplies are not endless and the burgeoning population already is placing heavy burdens on fresh water supplies in many areas.

In developing any new concept, unforeseen problems arise. Talented critics and obstructionists inevitably will make their appearance and assemble their followers. And if critics can cause delays, that opens the door to cost escalation which provides further ammunition to our opponents. A high value must be placed on the need to persevere. As for my personal position, I can promise you plenty of perseverance.

The challenge of large-scale nuclear desalting is great and the route looks long and arduous. But when we consider the tremendous potential for improving the lot of all mankind, I accept that challenge willingly. I trust you will do the same.

References

1a. Paper given at Symposium on Nuclear Desalination, Madrid, Spain, November 1968. Reprint is paper in its entirety.
1. James T. Ramey, USAEC, *Nuclear Reactors Applied to Water Desalting*. Third International Conference on Peaceful Uses of Atomic Energy, Geneva, September 7, 1964. Co-authors: James K. Carr and R. W. Ritzmann. USAEC Public Announcement S-16-64.
2. James T. Ramsey, USAEC, *Nuclear Energy—Potential for Desalting*, First International Symposium on Water Desalination, Washington, D.C., October 8, 1965. Co-authors: John A. Swartout and W. A. Williams, USAEC Public Announcement S-24-65.
3. James T. Ramey, USAEC, *Policy Considerations in Desalting and Energy Development and Utilization*, International Conference on Water for Peace, Washington, D.C., May 25, 1967. USAEC Public Announcement S-23-67.
4. Office of Science and Technology. Executive Office of the President, *An Assessment of Large Nuclear Powered Sea Water Distillation Plants*, and appendices, March 1964.
5. Interior Department-USAEC Report to the President, *Program for Advancing Desalting*, September 22, 1964.
6. *Nuclear Energy Centers, Industrial and Agro-Industrial Complexes, Summary Report*, Oak Ridge National Laboratory ORNL-4291), July 1968.
7. *Puerto Rican and U.S. Agencies Join In Study of Nuclear Centers*, USAEC Public Announcement L-257, November 8, 1968.

XIII

REVIEW AND ANALYSIS OF THE COSTS OF DESALTED SEAWATER[1]

Paul W. MacAvoy and Frederick J. Wells

Office of Water for Peace, U.S. Department of State, Washington

Summary of costs analyzed

Capital costs

ESTIMATES OF DESALINATION capital costs from various sources are summarized in Table 1 from Table 10. Capital costs, in dollars per daily gallon of productive capacity, are shown for the energy portion (steam and electricity), the desalination portion, and the total plant. The desalination portion of the total plant is that part of the plant where steam is used to desalt water. The energy section of the plant is where the steam and electricity are produced. For vapor-compression systems, the vapor compressor costs have been included in the desalination plant rather than in the energy plant.

Additional points concerning the derivation of Table 1 are listed below:

1) The energy plant portion of the costs for dual purpose plants producing water and electricity have been derived by using the available energy method or an approximation to this method based on estimated potential electrical output.

2) For the sake of simplicity, many items—e.g., the power to water ratio—relevant to the comparison of these costs, have not been included in Table 1.

3) No adjustment has been made to compensate for different dates of source estimates. Since there has been a general price inflation over the period, estimates prepared earlier will on this account have somewhat lower dollar costs than more recent estimates. Except for nuclear reactor costs, this is probably not a very significant source of error. Adjustment of costs to a

Table 1 Reported Capital Costs of Desalination ($ per Daily Gallon of Capacity)

Source of Estimate	Oak Ridge National Lab	Key West Plant	Jidda Estimate	ORNL	Struthers	Bechtel-OSW72	Allis-Chalmers	AREL	Aqua-Chem	Badger Co.	Bruns & Roe No.1	Burns & Roe No.2	Foster Wheeler	Fluor	Lock-heed	Parsons Co.	Worth-ington	B-L-H	Fergu-son
Plant Capacity	2.04 mgd	2.62	5.0	5.719	8.04	14.0	50.0	50.0	50.0	50.0	50.0	50.0	50.0	50.0	50.0	50.0	50.0	50.0	50.0
Desalination Process	VC-VTE-MSF	MSF	MSF	VC-VTE-MSF	VC-VTE-MSF	MSF	MSF	MSF	MSF	MSF	MSF	MSF	MSF	MSF	MSF	MSF	MSF	MEMS	HSME
Capital Costs Energy Plant	$0.34		0.34	0.21	0.21														
Desalting Plant	0.93		0.91	0.94	0.72	0.75	0.53	0.59	0.43	0.56	0.50	0.56	0.66	0.69	0.71	1.02	0.81	0.76	1.21
Total	$1.28	1.32	1.25	1.15	0.93														

Source of Estimate	Braun & Co.	Dow Chemical	GE	Average for the 50 mgd estimates	Kaiser Catalytic	Revised K-C	Parsons Co.	Parsons Co.	MWD	Revised MWD	Parsons Co.	Parsons Co.	ORNL	Parsons Co.	Parsons Co.	Mexico Study			
Plant Capacity	50.0 mgd	50.0	50.0		100.0	100.0	100.0	100.0	150.0	150.0	150.0	150.0	250.0	250.0	300.0	300.0	1000.0	1000.0	
Desalination Process	LTV-MSF	LTV	LTV-MSF		MSF (N)	MSF (N)	MSF (N)	MSF (N)	MSF (N)	MSF (N)	MSF (N)	MSF (N)	MSF (N)	VTE-MSF (N)	MSF (N)	MSF	MSF (N)	VTE-MSF (N)	
Capital Costs Energy Plant	$0.68				0.35	0.39	0.18	0.12	0.14	0.24	0.18	0.12			0.17	0.11	0.08	0.11	
Desalting Plant	0.79	0.55		0.69	1.10	0.98	0.70	0.67	0.67	0.87	0.69	0.67	0.41	0.36	0.67	0.66	0.54	0.30	
Total					1.45	1.37	0.88	0.79	0.82	1.11	0.86	0.79			0.84	0.77	0.62	0.41	

Notes: Site land or island, and product water conveyance costs excluded.

1. mgd = millions gallons per day
2. (N) = nuclear reactor energy source (otherwise fossil fuel)
3. VC = vapor compression
4. VTE = vertical tube evaporation
5. MSF = multistage flash evaporation
6. MEMS = multi-effects multistage flash evaporation
7. HMSE = horizontal submerged tube multiple effect evaporation
8. LTV = long tube verticle process

common base year has not been attempted because of uncertainties over the appropriate adjustment factors to use.*

4) Capital costs vary as a function of interest rate so that, for purposes of comparison, capital costs based on fixed charge rates of around five or six per cent have been used.

5) Land costs and product water conveyance costs have not been included because these two items are highly dependent upon the particular site involved. Both items may contribute significantly to the capital costs. For instance, land and product water costs make up about thirty per cent of the total MWD plant costs.

From the data in Table 1, total capital costs have been plotted in Figure 1, and the energy and water plant capital costs have been plotted in Figure 2. A number of observations concerning total capital costs in Figure 1 can be made:

1. The most "realistic" of these estimates for large scale plants based on the current MSF technology are the revised Kaiser and revised MWD estimates. Both estimates, especially the revised MWD estimate, are more

---

* The following information may be useful in providing a quantitative estimate of inflationary trends.

In Arthur D. Little, Inc., *Competition in the Nuclear Power Supply Industry*, Prepared for the U.S. Atomic Energy Commission and the U.S. Department of Justice, NYO-3853-1 (Washington: Government Printing Office, December 1958), p. 149, it is stated that by early 1968 nuclear reactor capital costs had risen 25–30% since a low point in mid-1966. The Joint U.S.-Mexico-IAEA study team points out that capital costs of nuclear power plants have risen as much as 40% since 1966 (pp. 60–61). (The authors of this Mexico Study have chosen to ignore these price increases, however.) Bechtel Corporation states in its Revised MWD Study that by 1968 nuclear reactors costs increased 40 to 45% since 1965 (p. 6). While these are large increase, they apply only to a small part (approximately 20%) of the total capital costs of a large desalination plant.

For the desalting portion of total desalination plant, capital costs have not risen as rapidly. Bechtel Corporation indicates on page 4 of its Revised MWD Study that the desalting plant costs rose from $101 million in 1965 to $131 million in 1968 despite major savings in copper tubing costs by the use of "surplus" government stocks. Bechtel Corporation in its evaluation of the ORNL 250 mgd VTE–MSF plant determines that the price increase for this desalting plant is about 6% for the period from 1965 to the third quarter of 1967 (p. 176 of OSW Report No. 391). Kaiser-Catalytic show a price decline of about 7.6% for desalting plant capital costs from 1965 to mid-1968. (Compare desalting plant capital costs on page 114 of the original Kaiser-Catalytic study with those in Table 2 of the 1978 report.) It is nuclear why such varying estimates of desalting plant investment cost trends should occur.

224 Desalting Seawater

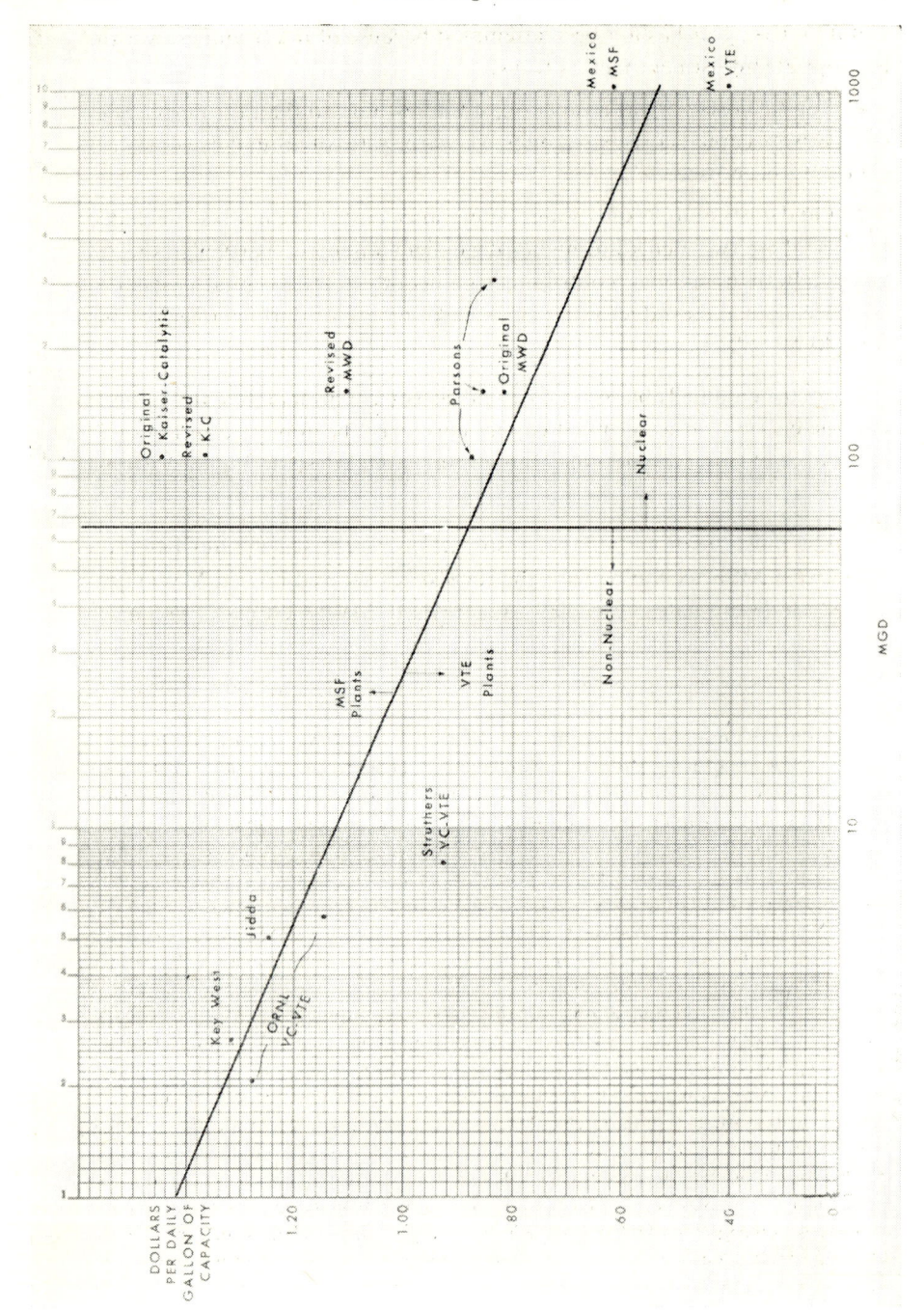

Figure 1 Total capital cost of desalination plants (Excluding Land and Product Water Conveyance)

The Costs of Desalted Seawater

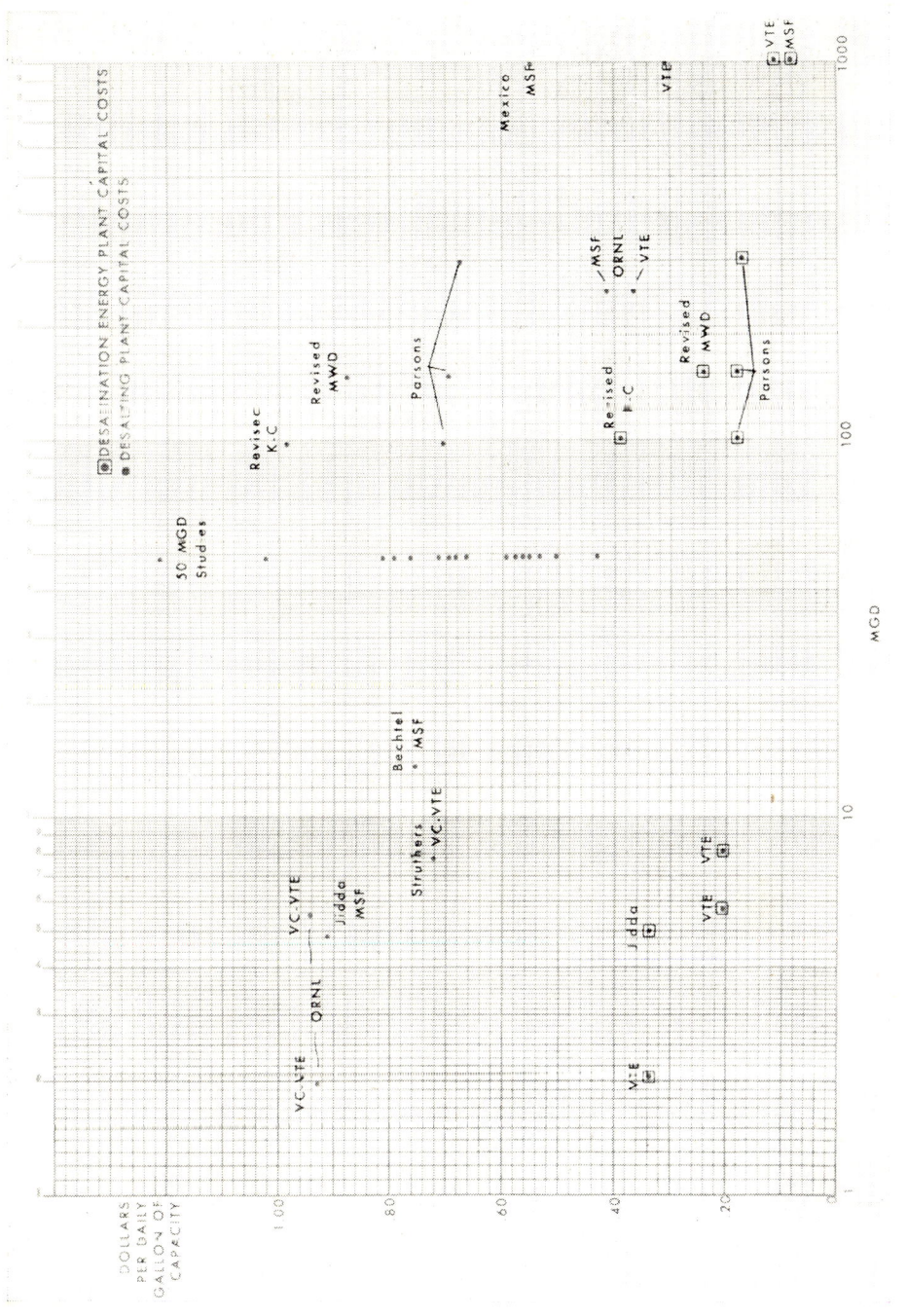

Figure 2 Energy and Desalting Plant Capital Costs (Excluding Land and Product Water Conveyance)

realistic than many of the others because contractual agreements for actual construction based on them were being considered. There is little danger in preparing optimistic cost estimates under OSW "study" contracts, but more caution must be shown in pre-bidding estimates such as the revised Kaiser and MWD estimates.

2. In relationship to the the actual Key West bid by Westinghouse in Figure 1, it is amazing how little decline in total unit capital costs occurs in the Kaiser and MWD estimates for much larger plants. In allocating capital costs between electricity and water, efforts were made to reduce the water capital costs at the expense of electric power capital costs. Thus, even the rather small difference in total capital costs between the Key West MSF plant and the revised MWD plant may be overstated.*

3. The Parsons Company estimates were, by their own admission, based on the original MWD capital costs and, therefore, are understated.

4. The Mexico Study of MSF costs appears highly optimistic. It is unlikely that scale economies could bring about such differences as shown between the revised MWD and the Mexico Study MSF plant since there is so little decline between the Key West MSF plant costs and the revised MWD plant costs.

5. With the possible exception of the two smallest plant estimates, the VTE estimates also seem optimistic. As yet the VTE technology has not been fully tested out. Even if VTE technology does perform up to expectations, there is still little reason to believe that such radical improvements in capital costs will result.

Observations about Figure 2 include:

1. As in Figure 1, the revised Kaiser and revised MWD estimates are probably the most reliable for large MSF plants. They are also among the highest set of estimates.

2. Except for two of the studies, the 50 mgd studies are all highly optimistic in comparison with the revised Kaiser and MWD estimates of desalting plant capital costs. Since price inflation for desalination equipment has not been very great, it is not possible to ascribe the difference to inflation.

3. The Mexico Study MSF and, to a greater degree, the ORNL MSF water plant costs appear too low to be reconciled with the revised MWD costs.

---

* On the other hand, the Key West plant capital costs are slightly understated because the capital costs for electricity production are not included.

The Costs of Desalted Seawater

4. Similar conclusions hold concerning MSF energy plant capital casts. The low Mexican Study MSF energy capital cost is a result of a conscious decision by the authors of that study to ignore the rise in reactor costs since 1965.*

5. VTE capital costs are estimated as being lower than MSF costs but the VTE technology has not yet been tested sufficiently for an accurate determination of the possible cost savings. Based on the differences in the two ORNL 250 mgd water plant costs, only a rather small saving in capital costs could be expected. A much larger difference between the MSF and VTE water plant capital costs is projected in the Mexico Study.

Unit product water costs

Costs of desalinated seawater in cents per thousand gallons are given in Table 2. They are based on the source costs provided in Table 10, but have been "normalized" as the following notes indicate:

1. Fixed charges on the capital costs of the energy portion of the plant and on the capital costs of the desalting portion of the plant were derived by applying a fixed charge rate of six percent to the capital costs in Table 1. A six percent fixed charge rate, which is composed of a 4.3% interest rate and a 30 year 1.7% sinking fund depreciation factor, was used since it approximates most of the fixed charge rates used in the various source estimates summarized here. An annual utilization or load factor rate of 90% was picked on the same basis.

It was desirable to pick a fixed charge rate utilization rate close to the rates used in the original estimates to avoid estimating changes in other parameters (such as the capital costs) which would result from changes in the fixed charge rate and the utilization factor. For cost estimates which supplied annual steam and electricity costs only, adjustment for different fixed charge and utilization rates would have been very difficult. In those cases where annual steam and electricity costs only were given, no adjustment was made because of different fixed charge rates and utilization rates used. Since these rates differed little from the sets used in the source cost estimates, the errors are not very large.

2. For consistency, insurance and interim replacement costs, both of which were added into fixed charges in some studies, were added to operations and

---

* Joint U.S.–Mexico–IAEA Study Team, p. *60*.

Table 2 Reported Unit Water Costs (in cents per 1000 gallons)

Source of Estimate	Oak Ridge National Lab	Jidda Study	ORNL	Struthers-Pratt & Whitney	Bechtel Corp. OSW No. 72	Allis-Chalmers	AREL	Aqua-Chem	Badger Co.	Burns & Roe No. 1	Burns & Roe No. 2	Foster-Wheeler	Fluor Corp.	Lockheed	Parsons Co.	Worthington Corp.	Baldwin-Lima-Hamilton	Ferguson
Plant Capacity	2.04 mgd	5.0	5.719	8.04	14.0	50.0	50.0	50.0	50.0	50.0	50.0	50.0	50.0	50.0	50.0	50.0	50.0	50.0
Desalination Process	VC-VTE-MSF	MSF	VC-VTE-MSF	VC-VTE-MSF	MSF	MSF	MSF	MSF	MSF	MSF	MSF	MSF	MSF	MSF	MSF	MSF	MEMS	HSME
Unit Water Costs																		
Energy Plant																		
Fixed Charges	6.3	6.1	3.9	3.9														
Fuel and Electricity	9.2	29.2	12.8	13.0														
O & M	6.3	1.7	4.5	2.4														
Subtotal	21.8	37.0	21.2	19.3	32.6	16.2	18.4	15.2	14.0	20.4	24.8	17.0	17.7	20.1	21.5	13.5	16.9	15.7
Desalting Plant																		
Fixed Charges	17.1	16.7	17.2	13.1	13.7	9.7	10.8	7.8	10.2	9.2	10.3	12.1	12.6	12.8	18.6	14.8	14.0	22.2
Chemicals	3.3	3.0	2.8	3.4	1.5	2.4	2.0	2.4	2.8	1.6	1.6	1.8	3.2	3.2	3.3	2.8	2.6	2.1
O & M	16.7	13.9	9.4	7.9	9.3	6.1	4.9	3.5	5.1	3.5	3.8	4.0	8.1	6.2	8.7	4.8	3.5	7.3
Subtotal	37.1	33.6	29.4	24.4	24.5	18.7	17.6	13.7	18.1	14.3	15.6	17.9	23.9	22.2	30.6	22.4	20.0	31.6
Total	58.8	70.7	50.6	43.8	57.1	34.9	36.0	28.9	32.1	34.7	40.4	34.9	41.6	42.3	52.1	35.9	36.9	47.3

Table 2 (continued)

Source of Estimate	Braun & Co.	Dow Chemical	General Electric Co.	Averages for the 50 mgd Plants			Kaiser-Catalytic	Revised Kaiser-Catalytic	Parsons Co		MWD Plant	Revised MWD Plant	Parsons Co.		Oak Ridge National Lab		Parsons Co.		Mexico Study	
				All	MSF and MEMS only	TLV-MSF only (Braun & G.E.)														
Plant Capacity	50.0	50.0	50.0				100.0	100.0	100.0	100.0	150.0	150.0	150.0	150.0	250.0	250.0	300.0	300.0	1000.0	1000.0
Desalination Process	LTV-MSF	LTV	LTV-MSF				MSF (N)	MSF (N)	MSF (N)	MSF	MSF (N)	MSF (N)	MSF (N)	MSF	MSF (N)	VTE-MSF (N)	MSF (N)	MSF	MSF (N)	VTE-MSF (N)
Unit Water Costs																				
Energy Plant																				
Fixed Charges							6.0	6.8	3.4	1.9	2.6	4.4	3.4	2.2			3.2	2.1	1.6	2.0
Fuel and Electricity							7.0	6.5	4.7	11.5	5.6	5.6	4.8	11.5			4.8	11.3		
O & M							1.8	1.9	0.9	0.3	1.3	1.3	0.9	0.6			0.9	0.4		
Subtotal	13.1	20.9	13.3	17.4	18.0	13.2	14.9	15.2	9.0	13.7	9.5	11.3	9.1	14.2	5.8	5.6	8.8	13.8	6.9	5.1
Desalting Plant																				
Fixed Charges	12.3	14.4	10.1	12.6	11.9	11.2	20.0	17.7	12.2	12.2	12.3	15.9	12.0	12.0	7.5	6.6	11.8	11.8	9.5	5.2
Chemicals	2.0	2.5	3.3	2.5	2.5	2.6	3.1	3.1	3.0	3.0	2.4	2.4	3.1	1.1	1.4	1.4	3.1	3.1	5.4	3.3
O & M	3.8	4.8	4.5	5.2	4.2	4.2	4.1	3.7	3.5	3.5	2.9	2.9	2.9	2.9	2.1	2.4	2.4	2.4		
Subtotal	18.1	21.8	17.9	20.3	19.6	18.0	27.2	24.5	18.7	18.7	17.6	20.2	18.0	18.0	11.0	10.3	17.3	17.3	14.9	8.5
Total	31.2	42.7	31.2	37.7	37.6	31.2	42.1	39.7	27.7	32.4	27.1	32.5	27.1	32.2	16.8	15.9	26.1	31.1	21.8	13.6

Notes

See Table 1 notes for desalination process abbreviations.
Costs have been normalized by applying the following parameters: A 6% fixed charge rate, a 90% plant load factor, and fossil fuel costs of 30¢ per million BTU

maintenance (O & M) costs. Nuclear insurance was also included in the O & M costs. O & M costs were also adjusted to reflect an annual utilization rate of 90%.

3. For fossil fueled plants, fuel costs have been standardized at 30 cents per million BTU.

4. The division of energy costs in dual purpose plants has been done on the basis of the available energy method or the closely equivalent potential electric power method.

5. Costs for land and product water conveyance have been left out since they depend upon the particular site location. In some cases these costs are an important percent of the total costs—e.g., about 20% in the case of the MWD plant.

6. No adjustment for price increase has been made. Such adjustments might tend to reduce part of the discrepancies in costs discussed next.

The data in Table 2 are plotted in Figures 3 through 5. The following conclusions are based on these data:

1. Much the same pattern of water costs prevails as in the case of capital costs.

2. For large, near-term MSF plants, the revised Kaiser and MWD estimates are probably the most acurate.*

3. The Parsons Company estimates (shown in Figure 3 only) are based on the original MWD estimates which are no longer valid. They are useful only to indicate the effects of scale on costs in the 100 to 300 mgd MSF plant range.

4. The ORNL MSF cost estimate is not compatible with the revised MWD estimate. It is also unlikely that the Mexico Study MSF estimate can be reconciled with the MWD estimate by economy of scale arguments. As already indicated, the reactor and power plant cost assumptions of $100 per electrical kilowatt of capacity used in this study should have been adjusted upwards to $150 or more for the late 1970 time period discussed.

5. Although in Figure 3 the 50 mgd plant cost estimates are probably compatible with the Kaiser and MWD estimates, there are some reasons to

---

* Revisions to annual non-fixed costs were very limited in these two studies. There was probably much less necessity to revise these costs than capital costs. For this reason the non-fixed charge costs are likely to be rather optimistic.

232 *Desalting Seawater*

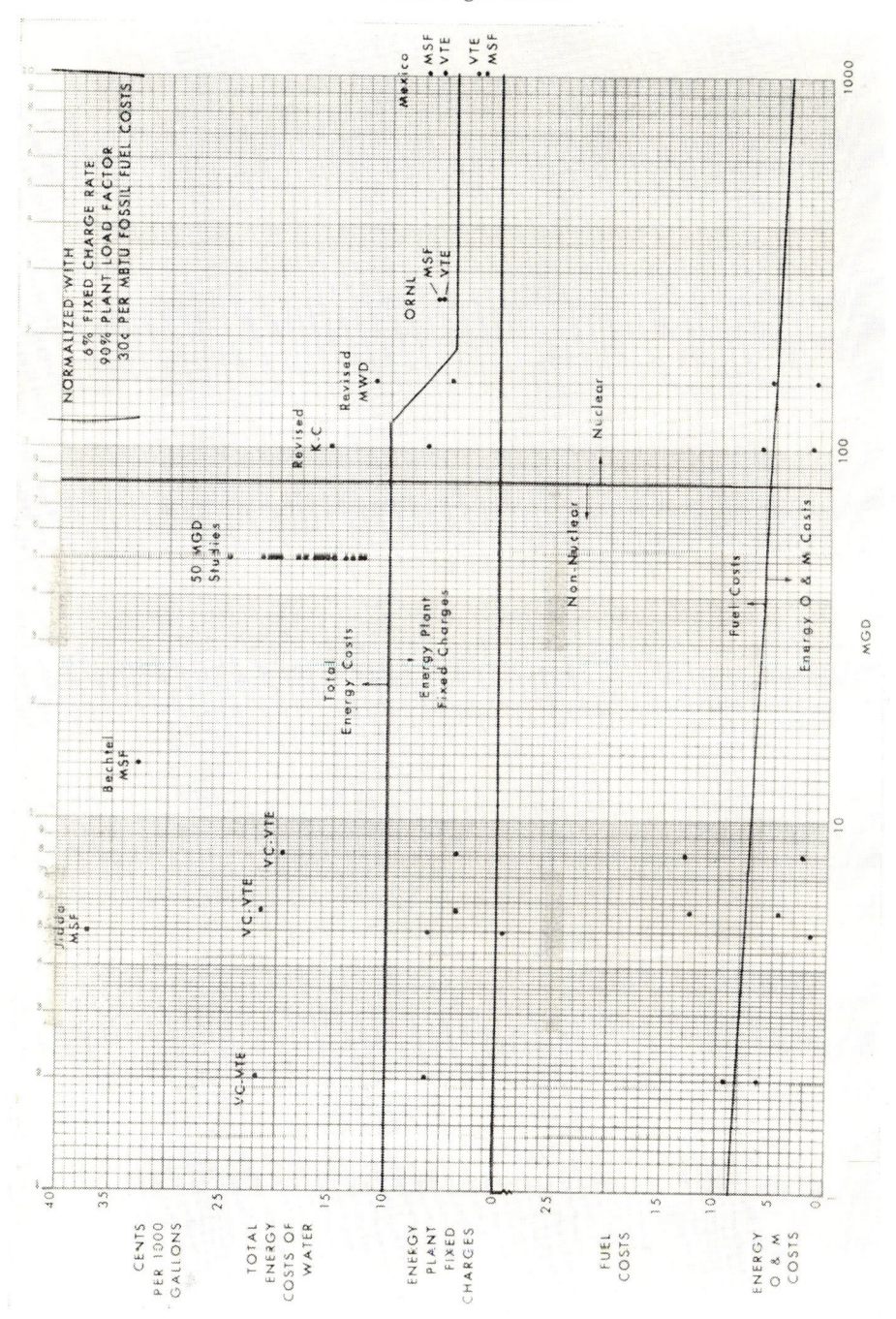

Figure 4 Normalized Energy Costs per 1000 Gallons.

The Costs of Desalted Seawater 231

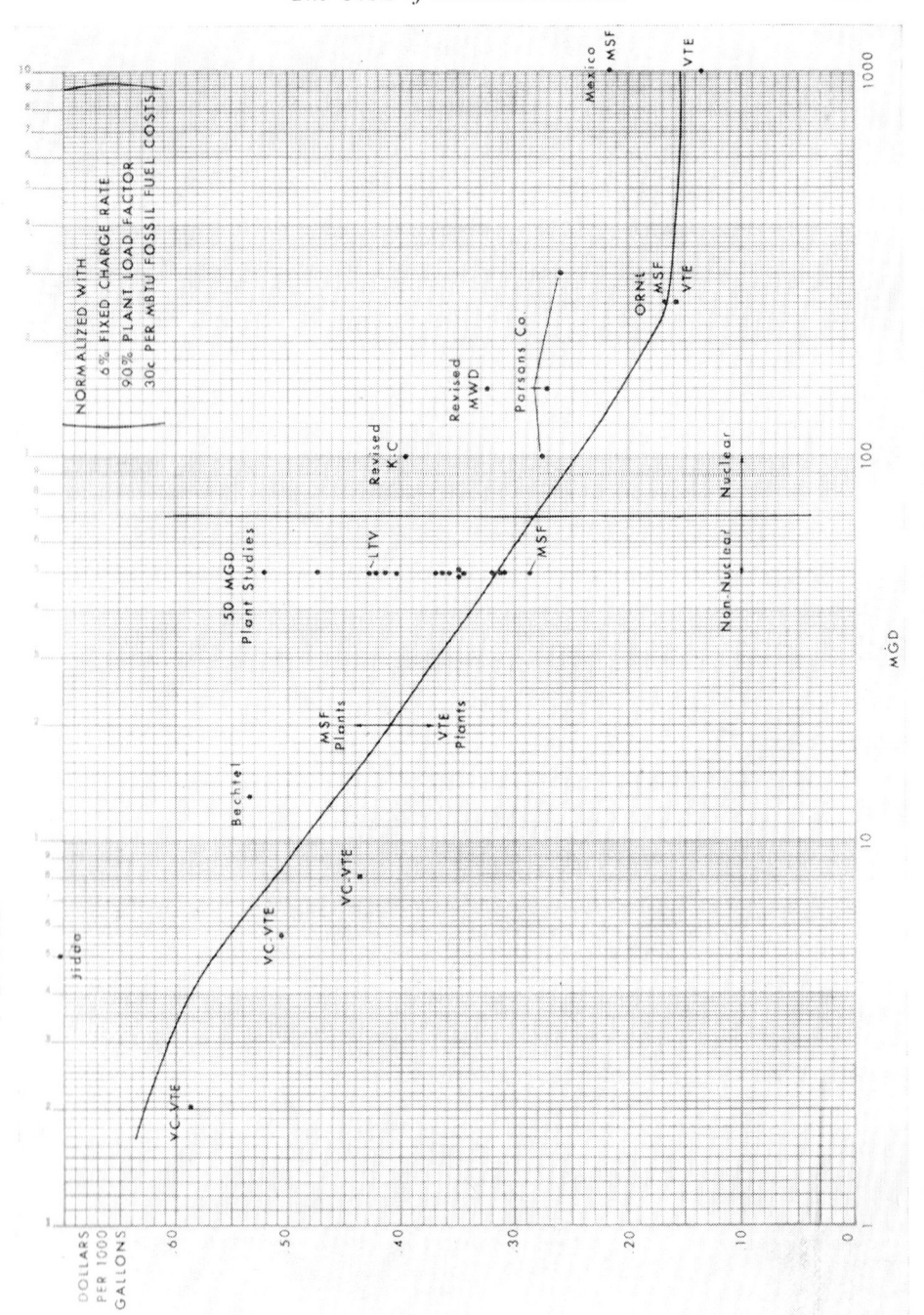

Figure 3 Total Unit Normalized Water Costs (Excluding Land and Product Water Conveyance).

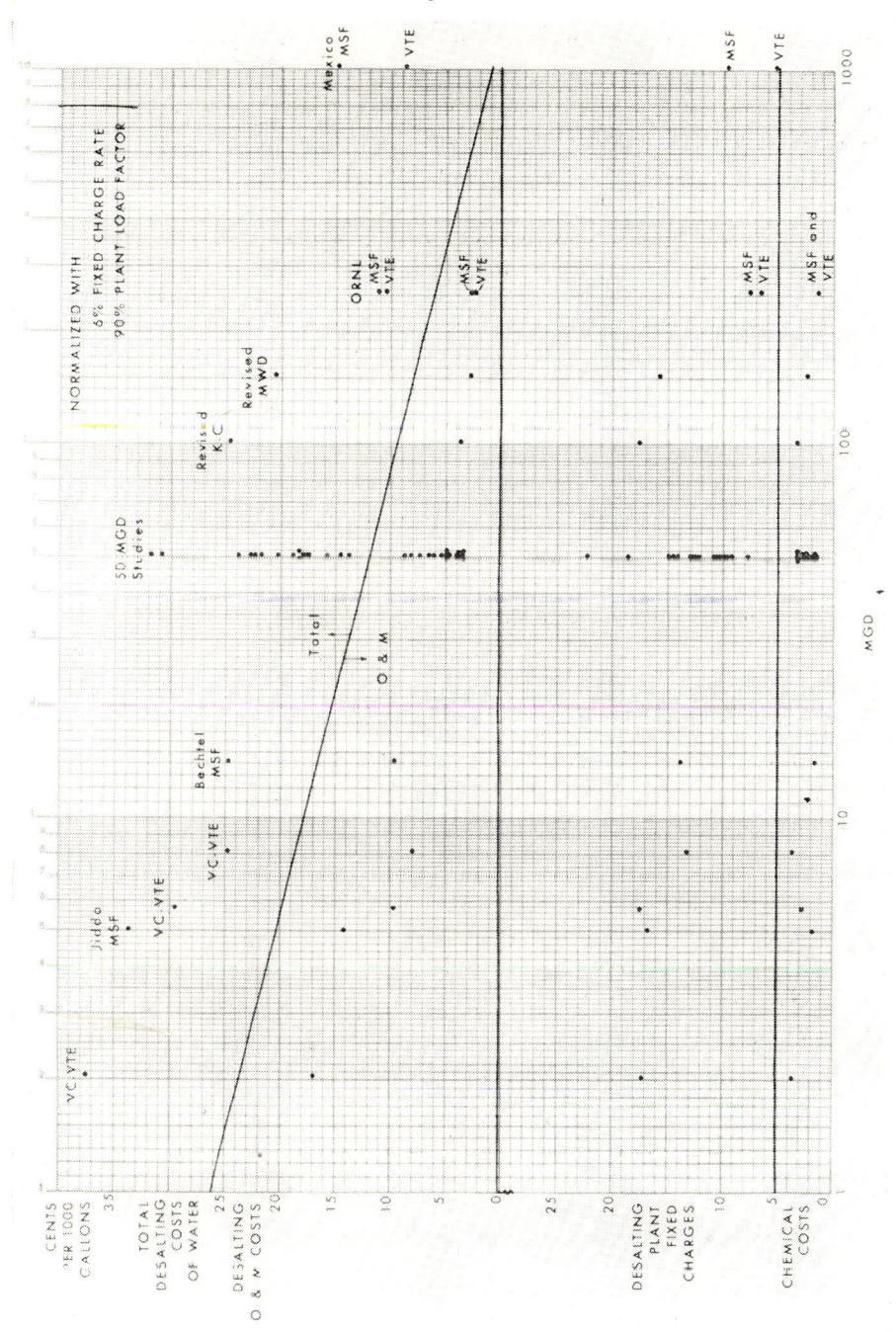

Figure 5 Normalized Desalting Unit Water Costs (Excluding Energy Costs).

believe that the 50 mgd costs are slightly too low relative to the revised Kaiser and MWD estimates. In Figure 5, the costs of the desalting portion of the process indicate that the 50 mgd study estimates are slightly optimistic relative to Kaiser and MWD. The differences in energy costs in Figure 4 are due to the type of energy source. If the 50 mgd plants had been based on nuclear reactor steam costs, then the resulting overall costs would have been lower and the optimistic bias present in most of these studies relative to MWD would be more apparent. As it is in Figure 3, the optimistic desalination non-energy costs are offset by the higher fossil fuel energy costs.

6. A serious deficiency of our study exists concerning the near-term costs of small MSF plants. The two estimates given—Jidda and Bechtel—are both "study" estimates. Actual costs of operation of recently built MSF plants such as the Key West plant and the San Diego demonstration plant should have been included. On the basis of one report, the actual cost of the 2.62 mgd Key West plant is close to one dollar per 1000 gallons if the effects of a "subsidized interest rate loan from the Federal government" are removed.* Although the plant load factor, fixed charge rate, fuel cost, and other items involved in this statement about Key West costs are unknown, it certainly appears that actual near-term costs of desalination in small MSF plants are significantly higher than shown.

7. Whether the indicated reductions in costs of MSF plants from economies in scale and dual purpose operations will result has yet to be proven. Even the reduction from the Jidda and the Bechtel level of costs to the Kaiser and MWD levels may not occur.

8. The VTE estimates are also based on an as yet unproven technology. Even if the technological advances expected from the VTE concepts are realized, the cost projections given may still be highly optimistic for both large and small sized plants. The low MSF cost estimates given by the Mexican Study group and particularly by ORNL tend to cast doubt on their VTE estimates.

---

* Marion Clawson, Hans H. Landsberg and Lyle T. Alexander, "Desalted Seawater for Agriculture: Is It Economics?", *Science*, **164** (June 6, 1969), p. 1145.

Bibliography of reference sources

50 million gallons a day conceptual design studies

Allis-Chalmers Manufacturing Company, *Conceptual Design of a 50 MGD Desalination Plant*, prepared for the U.S. Office of Saline Water, Washington, Government Printing Office, July 1965.

Applied Research and Engineering, Ltd., *Conceptual Design of a 50 MGD Desalination Plant*, prepared for the U.S. Office of Saline Water, Washington, Government Printing Office, 1965(?).

Aqua-Chem, Incorporated, *Conceptual Design of a 50 MGD Desalination Plant*, prepared for the U.S. Office of Saline Water, Washington, Government Printing Office, December 1965.

The Badger Company, *Conceptual Design of a 50 MGD Desalination Plant*, prepared for the U.S. Office of Saline Water, Washington, Government Printing Office, August 1965.

Baldwin–Lima–Hamilton Corp., *Conceptual Design of a 50 MGD Desalination Plant*, prepared for the U.S. Office of Saline Water, Washington, Government Printing Office, April 1966.

C. F. Braun and Comapny, *Conceptual Design of a 50 MGD Desalination Plant*, prepared for the U.S. Office of Saline Water, Washington, Government Printing Office, August 1965.

Burns and Roe, Inc., *Conceptual Design of a 50 MGD Desalination Plant*, prepared for the U.S. Office of Saline Water, Washington, Government Printing Office, October 1965.

Dow Chemical Company, *Conceptual Design for a 50 MGD Desalination Plant*, prepared for the U.S. Office of Saline Water, Washington, Government Printing Office, 1965(?).

The H. K. Ferguson Company, *Conceptual Design of a 50 MGD Desalination Plant*, prepared for the U.S. Office of Saline Water, Washington, Government Printing Office, August 1965.

The Fluor Corporation, Ltd., *Conceptual Design of a 50 MGD Desalination Plant*, prepared for the U.S. Office of Saline Water, Washington, Government Printing Office, August 1965.

Foster Wheeler Corporation, *Conceptual Design of a 50 MGD Desalination Plant*, prepared for the U.S. Office of Saline Water, Washington, Government Printing Office, September 1965.

General Electric Company, *Conceptual Design of a 50 MGD Desalination Plant*, prepared for the U.S. Office of Saline Water, Washington, Government Printing Office, November 1965.

Lockheed Missiles and Space Company, *Conceptual Design of a 50 MGD Desalination Plant*, prepared for the U.S. Office of Saline Water, Washington, Government Printing Office, 1965(?).

The Ralph M. Pearsons Company, *Conceptual Design of a 50 MGD Desalination Plant*, prepared for the U.S. Office of Saline Water, Washington, Government Printing Office, August 1965.

Worthington Corporation, *Conceptual Design of a 50 MGD Desalination Plant*, prepared for the U.S. Office of Saline Water, Washington, Government Printing Office, August 1965.

Other studies

Bechtel Corporation, *Engineering and Economic Feasibility Study for a Combination Nuclear Power and Desalting Plant: Summary*, prepared for the Metropolitan Water District of Southern California, the Office of Saline Water in the U.S. Department of the Interior, and the U.S. Atomic Energy Commission. December 1965.

Bechtel Corporation (?), *Cost Information—Bolsa Island Project*. 1968 (?). (Mimeographed.)

Bechtel Corporation, *A Study of Large Size Saline Water Conversion Plants*, prepared for the Office of Saline Water in the U.S. Department of the Interior. Office of Saline Water Research and Development Progress Report No. 72, Washington, March 1963.

Kaiser Engineers and Catalytic Construction Company, *Engineering Feasibility and Economic Study for (a) Dual Purpose Electric Power-Water Desalting Plant for Israel*, prepared for the U.S.-Israel Joint Board, Report No. 66-1-RE. Springfield, Virginia: Clearinghouse for Federal Scientific and Technical Information, January 1966.

Kaiser Engineers and Catalytic Construction Company, *Engineering Feasibility and Economic Study for (a) Dual-Purpose Electric Power—Water Desalting Plant: Effects on Costs of Increasing Capacity to 300 Megawatts*, prepared for the U.S.-Israel Joint Board. Report No. 67-35-RE. Oakland, California, Kaiser Engineers, September 1967.

Kaiser Engineers, *Review of Engineering Feasibility and Economic Study for (a) Israel Dual Purpose Electric Power—Water Desalting Plant: Estimates Based on Mid-1968 Costs and Escalation Based on Plant Design Commencing Mid-1969, with Construction Completion 1975*. Report No. 68-25-RE (Supplement), Oakland, California, Kaiser Engineers, October 1968.

Oak Ridge National Laboratory, *Conceptual Design Study of a 250 Million Gallons Per Day Combined Verticle Tube-Flash Evaporator Desalination Plant*, prepared for the Office of Saline Water, OSW Research and Development Progress Report No. 391, Washington, Government Printing Office, August 1968.

Oak Ridge National Laboratory, *The Oak Ridge National Laboratory Conceptual Design of a 250-MGD Desalination Plant*, prepared for the Office of Saline Water. OSW Research and Development Progress Report No. 214. Washington: Government Printing Office, September 1966.

Oak Ridge National Laboratory, *Preliminary Design of a Diesel-Powered Vapor-Compression Plant for Evaporation of Seawater*, prepared for the Office of Saline Water, OSW Research and Development Progress Report No. 276, Washington, Government Printing Office, August 1967.

The Ralph H. Parsons Company, *Engineering Study of the Potentialities and Possibilities of Desalting for Northern New Jersey and New York City*, prepared for the Office of Saline Water, OSW Research and Development Progress Report No. 207, Washington, Government Printing Office, April 1966.

Struthers Energy Systems, Inc. and Pratt and Whitney Aircraft, *Design and Economic Study of a Gas Turbine Powered Vapor Compression Plant for Evaporation of Seawater*, prepared for the Office of Saline Water, OSW Report No. 377. Washington, Government Printing Office, 1968 (?).

Technology Services, Inc., *Summary Evaluation of Conceptual Design for 50 MGD Desalination Plant*, prepared for the Office of Saline Water, OSW Research and Development Progress Report No. 277, Washington, Government Printing Office, August 1967.

U.S. Department of the Interior, *Preliminary Appraisal Report on (a) Combination Sea Water Desalting and Electric Power Plant for Jidda, Saudi Arabia*, prepared for the Government of Saudi Arabia, June 1964.

U.S.–Mexico–International Atomic Energy Agency study team, *Nuclear Power and Water Desalting Plants for Southwest United States and Northern Mexico: A Preliminary assessment*, TID-24767, Springfield, Virginia, Clearinghouse for Federal Scientific and Technical Information, September 1968.

Reference

1. Reprint is Summary and Bibliography.

XIV

DRY LANDS AND DESALTED WATER[1a]

Gale Young*

Desalination can supply water to create new living space; can it also create useful new farmland?

Men spread now, with the whole power of the race to aid them, into every available region of the earth. Their cities are no longer tethered to running water and the proximity of cultivation ... they lie out in the former deserts, those long-wasted sun baths of the race, they tower amidst eternal snows. ... One may live anywhere.
H. G. WELLS *The World Set Free—A Story of Mankind*[1].

Mounting population and pollution exert pressure on man to enlarge his living space. At present, the most rapidly growing part of the United States is the desert of the Southwest, into which the Colorado and Rio Grande rivers have been largely diverted. Water from the Feather River will shortly be added, crossing over a mountain range 2000 feet (600 meters) high, by way of conveyance facilities now under construction.

The building of the Colorado Aqueduct of the Metropolitan Water District of Southern California was undertaken 40 years ago. Since then, the assessed valuation of the region served by this water supply has increased by $20 billion, or by $5 for each 1000 gallons of water that has flowed through the pipes.

These remarks are not intended to impute some exaggerated value to water, but are made, rather, to point out that people seem to like desert climates (better than, say, rain forests or frozen tundras) and that they will move into them and build vigorous societies. Since the water is essential to the welfare of all, whether they be large direct users or not, the cost of the water in arid areas is often carried in part as overhead. For example, the Metropolitan Water District derives funds to meet about half of its costs from water revenues and levies taxes for the balance.

Water-supply projects tend to proceed by large steps. When they are first

---

* The author is assistant director of Oak Ridge National Laboratory, Oak Ridge, Tennessee.

completed there is an excess of water, and this can be used for agriculture while population and land values build up[2]. Later on, the suburban farm-land and its water allotments are absorbed in urban growth. This process is now under way around the large cities in our dry Southwest.

A third of the world's land is dry and virtually unoccupied, while half of the world's people are jammed—impoverished and undernourished—into a tenth of the land area. A major part of the coming increase in population will occur in the less-developed countries and, as *Nature* expressed it[3], "this huge army of uneducated, untrained, underfed, underprivileged recruits to humanity is being bred at the very moment when science and technology are rapidly undermining the need for and the status of the unskilled. ... What stands out most emphatically is in fact the insufficiency of economic factors or motives if the challenge is to be met".

The past decade has seen a quickening of interest in the large-scale desalt-ing of seawater as another means of opening up dry areas of the earth for human occupancy. Thus, Jacob Bronowski writes[4], "let me ask first what is going to be the single greatest technological change in the physical sciences over the next twenty or thirty years. My guess is that desalting of seawater is going to be the most important advance for overall world development. With-out this the whole complex program of bringing under-developed countries to an acceptable level of economics, education and political maturity is in-soluble." And Charles Lowe writes[5], "The desert is man's future land bank. Fortunately, it is a large one, offering eight million square miles of space for human occupation. It is also fortunate that it is a wondrously rich bank, which may turn green when man someday taps distilled seawater for irriga-tion. Bridging the gap from sea to desert will be greatly facilitated by the geographical nearness of most of the world's deserts to the oceans. When this occurs it will surely be one of the greatest transformations made by man in his persistent and successful role un changing the face of the planet". As an example, imagine what changes would follow if Baja California were to be opened up for settlement with the help of desalting and power station. What might another thousand miles of southern California be worth in a few generations?

This process has already begun to take place on a small scale in certain locations. Thus, Kuwait on the Persian Gulf and Shevchenko on the barren eastern shore of the Caspian Sea, where the Russians are building the world's largest desalting unit, are two oil field communities which depend upon desalination for their water supply. As in other locations where desalination would be employed, no cheaper sources of fresh water are available.

In general, desalination will be able to meet the freshwater needs of municipalities and industries located along the ocean shore, or in some interior regions having salty lakes or groundwater. And anywhere that cities go, some agriculture, for fresh fruits and vegetables at least, tends to follow. For example, Kuwait, in one of the most devilishly hot and sandridden environments on earth, has recently ordered acres of enclosed greenhouses at a cost of many thousands of dollars per acre to supply garden produce which will be fresher and reportedly cheaper than imported produce. But it is less clear whether agriculture that depends on desalted water ("desalination agriculture") may someday expand to include substantial production of staple foods as well. Here opinions differ, and, as the saying goes, the less light sometimes generates the more heat.

Water requirements for crops

In enclosed agriculture, such as Kuwait will have, so little fresh water escapes that its cost is not important. The situation is quite different in openfield agriculture, since the amount of water lost to the atmosphere in evapotranspiration is generally[6] many times that actually involved in the plant's growth reactions, and the cost of supplying this total by desalination is of prime importance.

Some authors have stated that the cost of desalted water will be for many years "at least one whole order of magnitude greater than the value of the water agriculture"[7]. Other writers have been more optimistic. Thus, the value of distilled water for agriculture in Israel has been estimated by MacAvoy and Peterson[8] to be about 27 cents per 1000 gallons. A generalized study[9] (p. 28) indicates that "arid tropical and semitropical regions with year-round growing climates appear potentially capable of growing food at costs in or near the world market import price range, using desalted water at prices like 20 cents per thousand gallons". Estimates have been made of the total (direct plus indirect) benefits of irrigation water in several U.S. locations. There is considerable variation, estimates of losses due to taking water away from crops being 25 to 36 cents per 1000 gallons for the Texas High Plains and 32 to 44 cents for the Imperial Valley[10].

As noted above, the growing of some high-value crops tends to follow the occupation of a dry area, and an economic enterprise would, of course, first seek to saturate the market for this produce. But one is not going to feed the teeming millions on orchids or avocados, so the inquiry turns inevitably to the bulk staple foods. Of the staples, the grains are the most important, and of the grains, rice is the staple and preferred food of half the human race. Rice, because of this demand, has a value double that of wheat on the world market—but it is not usually talked about in connection with desalination agriculture. Let us, therefore, talk about it.

Rice is indigenous to southern and eastern Asia, where much of the world's starvation occurs. In the monsoon countries, where fields are flooded periodically, the rice plant has the great virtue of being able to conduct oxygen to its roots through a hollow stem, and to flourish where other crops suffocate. Cultivation under such conditions leads to heavy use of water. In Table 1, from a talk by the dean of agriculture of the University of Sydney[11], it may be seen that in Australia, the most favorable of the instances shown, a price for water of 20 cents per 1000 gallons would add 10.6 cents per pound to the cost of rice. This is more than the rice is worth and provides little basis for enthusiasm about desalination.

Table 1 Amount of water used in growing rice in various countries.

Location	Pounds of water per pound of paddy rice	Gallons of water used per pound of milled rice
Thailand	10,000	1800
India	10,000	1800
Japan	4,900	900
California	3,300	610
Australia	2,900	530

Table 2 gives data on yield and water use for several other crops which are widely used. The "Rice" row of Table 2 is left blank for the present. The water-use rates were estimated (and checked against measured rates for farms when these were available) for a coastal site in the southeastern Mediterranean area, an irrigation efficiency (the fraction of the applied water which is lost to the atmosphere in evapotranspiration) of 80 percent being assumed. The other 20 percent of the water is that part which evaporates before reaching the crop, or which is carried away by deep percolation below the root zone. With efficient sprinkler or soaker irrigation systems this efficiency can usually be bettered, even without allowing for partial recovery of the deep drainage water. The yields assumed are those obtained regularly today by efficient farmers in production areas specializing in the crops in question. Record yields are considerably higher—more than double in the case of wheat, potato, tomato, and maize and in the case of an assumed yield for rice, discussed below.

Table 2 Relation of yield to water use for various crops.

Crop	Yield (10^3 lb/acre)	Food value (10^2 Cal/lb)	Water use Inches	Water use Gal/lb	Water use Gal/10^3 Cal
Grain					
Wheat	6.0	14.8	20.0	91	61
Sorghum	8.0	15.1	27.6	94	62
Maize	9.0	15.8	27.6	83	53
Rice					
Vegetable					
Potato	48	2.79	16.0	9.0	32
Tomato	60	0.95	19.0	8.6	91
Citrus					
Orange	44	1.31	53.1	33	250

Table 1 shows that rice growers typically use several times as much water per pound of grain as is needed for the other grains of Table 2. While some people may have doubted that rice requires this much water, this is the way the picture has stood historically. Just recently, however, results have appeared which do not support the classical view of rice as a water hog. In an experiment with IR 8 rice at India's Central Rice Institute, in Cuttack, a good crop [6350 pounds (2900 kilograms) per acre as paddy rice] was obtained with consumption of only 16.7 inches (42 centimeters) of water, irrigation being applied only when the soil was completely crusted. This work was reviewed and reported by J. S. Kanwar, deputy director general of the Indian Council of Agricultural Research[12]. Kanwar states, "The common notion that continuous submergence is essential for paddy is belied". In a corresponding experiment based on the classical practice of continuous submergence, six times as much water (100 inches) was used and the yield was 6 percent higher. Obviously, most of the 100 inches of water was forced into the subsoil by the head of standing water.

If confirmed in subsequent tests and field experience, the finding may turn out to be one of the landmarks in the war against hunger. We add that rice can be grown the year round in warm climates; for example, in the Philippines three crops per year have been grown, yielding a total of 18,000 pounds of rice per acre.

If the results of the Indian experiment are taken at face value, the missing numbers for rice, which can now be added to Table 2, are as follows: yield, 4.2 (10^3 pounds per acre); food value, 16.5 (10^2 Calories per pound); water use—16.7 inches, 108 gallons per pound, 65 gallons per 10^3 Calories. This puts rice right in among the other grains as a user of water. It may be seen that to supply a person a minimum adequate allowance of 2500 Calories per day[13] in this highly intensive and scientifically managed type of agriculture would require an average of 160 gallons of water per day in the case of grain, or half this in the case of potatoes.

Production cost estimates for grain

For any selected price of water, Table 2 enables one to compute the cost of water per pound of product, as illustrated for rice in Table 1. However, water is only one component of the total cost. We have, therefore, made some illustrative estimates of the total production costs for grain for a large, intensively operated farm. These costs are not expected to vary greatly with farm size. The estimates were made for a farm of 300,000 acres (120,000 hectares) which uses about a billion gallons of water per day.

These estimates are summarized in Tables 3 to 5. Table 3 shows the capital investment per acre, while Table 4 gives annual out-of-pocket costs for items which cannot be allocated directly to specific crops in the year-round rotation. The costs which can be specifically assigned are shown in Table 5.

Table 3 Estimated capital investment.

Item	Dollars per acre
Irrigation and storage-well system	375
Land, land development, and roads	85
Drainage system	110
Water for initial leaching	130
Grain storage	85
Machinery	115
Farm buildings	25
Interest during construction	25
Total	950

Table 4 Estimated overhead costs.

Item	Dollars per acre per year
Maintenance	14
Pumping power	1
Water losses in storage and canal leakage	12
Experimental station	1
Management and miscellancous	7
Total	35

Table 5 Estimated direct costs for grain production.*

Item	Dollars per acre per crop
Fertilizer	11
Pumping power	6
Seed	5
Labor	2
Machine operation	2
Storage and marketing	3
Other chemicals	3
Miscellaneous	8
Total	40

* Exclusive of the cost of irrigation water.

It should be noted that the water is priced at the farm inlet and that it is supplied to the farm at a rate that is constant throughout the year, except for the time when the desalting plant is shut down for maintenance or other reasons. Thus, because of seasonal variations in the requirement for irrigation water, it is necessary to resort to such measures as (i) making seasonal adjustments in the size of the irrigated area and (ii) storing water in underground aquifers at certain times of the year and pumping it back up at other times (10 percent of the stored water is assumed to be lost in this process). Allowance for such storage has been made in Table 3 and 4[14].

An interest rate of 10 percent on the investment shown in Table 3 amounts to $95 per acre per year, and allowance for depreciation and for working-capital needs adds about $10 more. With the overhead costs shown in

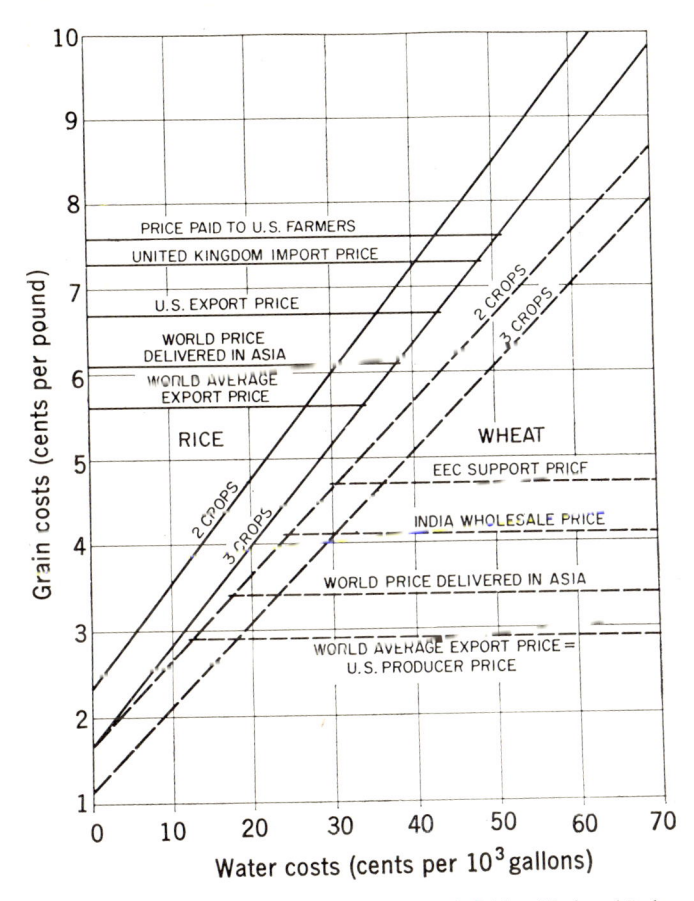

Figure 1 The costs of growing grain (see [30]). [Oak Ridge National Laboratory].

Table 4, the total fixed charges are thus $140 per acre per year. In regions where a second major crop, of the same or some other product, can be grown in the same year on the same land and share fixed costs with the grain, the cost of an acre crop of grain would be ($140/2) + $40 + cost of water = $110 + cost of water. In some areas, such as parts of India, where it might be possible to grow three crops per year the corresponding cost would be $87 + cost of water.

The resulting costs for milled rice and for wheat are plotted in Fig. 1, along with some representative prices.

In effect, what is here being contemplated is that the high cost of water may be at least partially offset by the opportunity to conduct intensive year-round "food factory" agriculture in favorable growing climates with many conditions under unusually good control. For example, recent observations suggest that very frequent or drip irrigation can cause substantial increases in the crop yield[15]. In these experiments the nitrogen fertilizer was applied in the water. There is also speculation that slow release of carbon dioxide beneath the plant canopy might enhance yields. Thus it is entirely possible that future yields may exceed those of Table 2.

Desalination

The newest United States desalting plant, at Key West, Florida, produces 2.5 million gallons of fresh water per day at a cost of about $1 per 1000 gallons. It is anticipated that the larger plants envisaged for future regional water supply will be able to attain lower costs through the economies of scale and the benefits of more advanced technology. A plant producing 8 million gallons per day[16], which will be built around a gas turbine, will have vapor-compressor heat pumps coupled to vertical-tube evaporators and will use the engine-rejected heat in flash evaporators, is expected to produce fresh water at a cost of about half that for the Key West plant[17].

Heat transfer enhancement is currently being introduced into evaporator designs. For example, doubly finned or doubly fluted vertical tubes give several times the heat transfer of plain smooth tubes, and their use can considerably reduce the size and cost of a plant[18]. To some degree the heat transfer in the horizontal tubes of a flash evaporator can be similarly augmented. Such advances are now being tried in test units and pilot plants. The preferred large-plant design at present is a combined flash and vertical-tube evaporator with heat transfer enhancement[19].

Economies can be achieved through the design and construction of dual-purpose plants which produce both power and water, as compared to single-purpose plants that produce only water. In terms of 1967 dollars at 10 percent interest[20], the cost of water at a large (billion gallons per day) single-purpose plant, in which most of the steam bypasses the turbine without generating any power, was estimated[9] (pp. 3, 4, 19, 20)[21] (p. 39) to be 26 to 32 cents per 1000 gallons, while at a dual-purpose station the range for incremental cost of water was 16 to 24 cents. At an interest rate of 9 percent, another study[22] indicated 15 to 26 cents for the dual-purpose cost range. These large-plant studies were based on the use of reactor heat sources, with technology ranging from reactors of the type (light-water) and size being built in the

United States today to advanced fast and thermal breeder reactors of the type that will be operational some two decades hence. Since these estimates were made there have been sharp increases in reactor prices, and it may turn out that the costs of water will have to be revised upward by 2 or 3 cents per 1000 gallons[23]. On the other hand, advanced evaporator and agricultural technologies appear to be moving ahead more rapidly than had been expected.

The saving effected through dual-purpose operation works the other way as well. Once you have decided to build a single-purpose steam bypass water plant, the incremental cost of producing power also is very low—say, 1.5 mills per kilowatt-hour[21] (p. 39) for reactor stations of the largest sizes being constructed today, such as Brown's Ferry (3000 electrical megawatts). This power may be transmitted to a network in some instances, or to adjacent industrial plants in a so-called agro-industrial complex, to help develop and build up the region. A number of industrial processes have been studied in this connection[24], as well as several possible locales[25] and some of the implementation problems[26]. One of the large energy consumers is the production of ammonia for fertilizer by way of electrolytic hydrogen, with only air and water used as raw materials. Potassium fertilizer can be produced from the sea, as can other products, such as chlorine, magnesium, salt, and caustic. Hoyle wrote[27]:

... the older established industries of Europe and America ... grew up around specialized mineral deposits—coal, oil, metallic ores. Without these deposits the older style of industrialization was completely impossible. On the political and economic fronts, the world became divided into "haves" and "have nots", depending whereabouts on the earth's surface these specialized deposits happen to be situated. ... In the second phase of industrialism, ... no specialized deposits are needed at all. The key to this second phase lies in the possession of an effectively unlimited source of energy. ... Low-grade ores can be smelted—and there is an ample supply of such ores to be found everywhere. Carbon can be taken from inorganic compounds, nitrogen from the air, a whole vast range of chemicals from seawater. So ... a phase in which nothing is needed but the commonest materials—water, air and fairly common rocks. This was a phase that can be practiced by anybody, by any nation. ... This second phase was clearly enormously more effective and powerful than the first.

Summary

The fastest-growing region in our country is the desert of the Southwest. This suggests that the vast, warm, dry areas of the world may be attractive for human occupancy as earth's population soars, if the water and power needs can be met. It may also be that nuclear energy—not tied by any umbilical cord to fossil deposits—will have a significant role in opening up these arid areas and creating usable land for human living space.

Since much arid land lies relatively near the sea and the aggregate length of the coastal deserts nearly equals the circumference of the globe, desalination is a freshwater source of broad potential applicability when cheaper alternative sources are not available. Cities and industries can now spread widely along the ocean shore, bringing with them some agriculture for garden produce.

A more difficult question concerns the extent to which desalination agriculture will be used in newly occupied arid lands for the production of staple foods. Since we cannot predict with any accuracy either the population growth or the increase in food production by various means, let alone the difference between these two large quantities, our knowledge of future food shortages is very poor indeed. It therefore appears prudent to conduct research, development, and pilot planting projects to investigate such potential methods for augmenting food production. These activities should be conducted on a scale sufficient to permit practical evaluation, so that the option of invoking such methods will be open, should circumstances require it. To quote R. R. R. Brooks[28], "The key to [the] ... future of two-thirds of the human species is rising productivity in agriculture. All political dogmas, party slogans, planning strategies, and models of economic growth shrivel to irrelevance in the face of this fact."

A preliminary and generalized study was conducted at the Oak Ridge National Laboratory with the collaboration of outstanding agricultural and engineering people from other countries and from U.S. government agencies, universities, foundations, and industries. This study considered intensive year-round farming in warm coastal deserts, based on the use of distilled seawater, in association with clustering industries. As indicated in Fig. 1, production costs for rice and wheat appear[29] to fall, for water costs of around 35 and 20 cents per 1000 gallons, respectively, in the general area of recent grain prices. These water prices, in turn, fall (barely) within the estimated future cost range for desalinated water.

The significant conclusion, we believe, is that desalination agriculture is in the realm of practical possibility, rather than being far afield. To our mind this appears of sufficient importance, in view of the population expansion and the interest in opening up new lands and communities, to warrant the development of advanced desalting plants and intensive scientifically managed agriculture. We hope that desert research farms running on distilled water, and controlled-environment agricultural test chambers, may become as well known to you on your TV screen in a few years as starving Biafran babies are today.

References and notes

1a. *Science*, January 23, 1970. Reprint is article in its entirety. Copyright by the American Association for the Advancement of Science.
1. H. G. Wells, *The World Set Free—A Story of Mankind* (Dutton, New York, 1915), p. 241.
2. In Los Angeles some strawberry growers use land worth *$10,000* per acre and pay city taxes on it.
3. *Nature* **199**, 411 (1963).
4. J. Bronowski, *Saturday Rev.* **1969**, 45 (5 July 1969).
5. C. H. Lowe, Jr., in A. S. Leopold, *The Desert* (Life Nature Library, New York, 1961), Introduction.
6. There are xerophytic plants (such as pineapple) that carry on photosynthesis with closed stomata in the daytime, using stored carbon dioxide absorbed through open stomata at night. They thus transpire much less water per unit of plant matter produced than ordinary crops do.
7. M. Clawson, H. H. Landsberg, L. T. Alexander, *Science* **164**, 1141 (1969).
8. P. W. MacAvoy and D. F. Peterson, Jr., *The Engineering Economics of Large-Scale Desalting in the 1970's* (Praeger, New York, in press).
9. "Nuclear Energy Centers; Industrial and Agro-Industrial Complexes: Summary Report", *Oak Ridge Nat. Lab. Rep. ORNL-4291* (1968).
10. C. W. Howe, in *Arid Lands in Perspective*, W. G. McGinnies and B. J. Goldman, Eds. (AAAS, Washington, D.C., 1969), p. 379.
11. J. R. A. McMillan, "Water, Agricultural Production and World Population" (Fatter Oration) (1965).
12. J. S. Kanwar, "From protective to productive irrigation", *Econ. Polit. Weekly Rev. Agr.* (29 March 1969). Data, cited from M. S. Chaudhury and R. G. Pandrey, are given in more detail in an article in *Indian Farming*, in press.
13. R. Revelle, *Proc. Nat. Acad. Sci. U.S.* **56**, 328 (1966).
14. If suitable aquifers are not present and an above-ground storage reservoir is constructed in the farming region, there is an increase in cost such that the value $140 in the next paragraph of the text changes to about $150 per acre per year, it being assumed that the reservoir is of a size adequate to supply a farm region with an intake of about 10^9 gallons of water per day.
15. D. Goldberg, B. Gornat, M. Shmueli, "Advances in irrigation in Israel's agriculture", paper presented at the 1st World Congress of Engineers and Architects, Israel (1967). See also D. Goldberg and M. Shmueli, "Drip irrigation—a method used under arid and desert conditions of high water and soil salmity", unpublished.
16. This plant is now in the preliminary planning stage.
17. "Design and Economic Study of a Gas Turbine Powered Vapor Compression Plant

for Evaporation of Seawater", *Office Saline Water Res. Develop. Progr. Rep. No. 377* (1968).
18. Such tubes are to be used in the gas turbine unit discussed in the preceding paragraph.
19. "Conceptual Design Study of a 250 Million Gallon per Day Combined Vertical Tube-Flash Evaporator Desalination Plant", *Office Saline Water Res. Develop. Progr. Rep. No. 391* (1968).
20. The cost of desalted water is sensitive to changes in interest rates because, like conventional means of providing new water supplies such as dams and aqueducts, the process requires a large capital outlay.
21. *Oak Ridge Nat. Lab. Rep. No. ORNL-4290* (1968).
22. "Nuclear Power and Water Desalting Plants for Southwest United States and Northwest Mexico" (1968), pp. 113 and 136; this is a preliminary assessment made by a joint United States-Mexico International Atomic Energy Agency study team.
23. This expectation is based on the assumption that combined flash and vertical-tube evaporators are adopted; otherwise, the increase in water costs would be expected to be larger.
24. A. M. Squires, "Steel making in an agro-industrial complex", in *Oak Ridge Nat. Lab. Rep. No. ORNL-4294* (1968); W. E. Lobo, "Acetylene production from naphtha by electric arc and by partial combustion", in *ibid,;* H. E. Goeller, "Tables for computing manufacturing costs of industrial products in an agro-industrial complex", *Oak Ridge Nat. Lab. Rep. No. ORNL-4296* (1969).
25. T. Tamura and W. J. Young, "Data Obtained on Several Possible Locales for the Agro-Industrial Complex", *Oak Ridge Nat. Lab. Rep. No. ORNL-4293* (1970).
26. J. A. Ritchey, "Problems in Implementation of an Agro-Industrial Complex", *Oak Ridge Nat. Lab. Rep. No. ORNL-4295* (1969); R. L. Meier, *Bull. At. Sci.* **1969**, 16 (March 1969); R. D. Sharma, *ibid.* **1969**, 31 (November 1969).
27. F. Hoyle, *Ossian's Ride* (Harper, New York, 1959), p. 146.
28. R. R. R. Brooks, *Saturday Rev.* **1969**, 14 (9 August 1969).
29. The estimates are based on an interest rate of 10 percent.
30. Figure 1 is based on data and publications as follows. The average price paid the U.S. producer for wheat in the period 1956–65; the average price per pound of milled rice, at 0.65 paddy weight (0.65 pound of milled rice from each pound of paddy), paid the U.S. producer in the period 1950-65; the average price of Siam Patna No. 2 rice imported into the United Kingdom in the period 1955–65; the average wholesale price of wheat in India (the Punjab) in the period 1951–65 *[FAO (Food Agr. Organ. UN) Prod. Yearb.* **20** (1966)]; the world average export value of wheat in the period 1961–67; the world average export value of rice in the period 1955–67 [*The State of Food and Agriculture: 1968* (Food and Agriculture Organization of the United Nations, Rome, 1968)]; the average U.S. export value of rice in the period 1960–65; the European Economic Community's wheat support price [*The World Food Problem* (President's Science Advisory Committee, Washington, D.C., 1967), vol. 2, pp. 145, 149]; recent world prices for wheat and rice delivered Asia, as given by R. P. Hammond in a paper presented at the International Conference on Water for Peace, Washington, D.C., May 1967.
31. The research discussed here is sponsored, by the U.S. Atomic Energy Commission under contract with the Union Carbide Corp.

XV

DESALTED SEAWATER FOR AGRICULTURE: IS IT ECONOMIC?[1a]

Marion Clawson, Hans H. Landsberg, and Lyle T. Alexander

In the past decade, there has been mounting advocacy of desalting seawater for use in commercial agriculture in various locations of the world, especially in the Middle East. The process, it is contended, is both technically and economically feasible, or soon will be, and its application on a large scale can produce additional volumes of food at competitive proces—the desert will blossom like the rose—and at a profit.

Although research on desalting techniques had proceeded for many years in the Interior Department's Office of Saline Water supported by modest funds, and the Atomic Energy Commission had been exploring the role that nuclear energy might play in desalting, the entire matter suddenly acquired international interest after the 6-day Israeli-Arab war in June 1967.

Within days after the war, the London *Times* published two letters recommending desalting schemes in the Middle East. A detailed letter from Edmund de Rothschild suggested three nuclear desalting installations in Israel, Jordan, and the Gaza Strip, respectively. This provoked comments and questions in the House of Commons, which generally approved the idea or at least further exploration of it. On this side of the Atlantic the U.S. Senate in December 1967 passed Resolution 155 without a dissenting vote. It says in part:

Whereas the greatest bar to a long-term settlement of the differences between the Arab and Israeli people is the chronic shortage of fresh water, useful work, and an adequate food supply; and

Whereas the United States now has available the technology and the resources to alleviate these shortages and to provide a base for peaceful cooperation between the countries involved:

Now, therefore, be it

Resolved, that it is the sense of the Senate that the prompt design, construction, and operation of nuclear desalting plants will provide large quantities of fresh water to both Arab and Israeli territories and, thereby, will result in

1) new jobs for many refuges;
2) an enormous increase in the agricultural productivity of existing wastelands;

3) a broad base for cooperation between the Israeli and Arab governments; and

4) a further demonstration of the United States efforts to find peaceful solutions to areas of conflict...

The resolution was a direct descendant of the "Strauss-Eisenhower Plan", a proposal by former AEC Chairman Lewis Strauss for which he obtained Eisenhower's backing. The proposal gained its greatest popularity through an article written by Eisenhower[1], which, with reference to the Middle East, states the proposition in its most optimistic form:

> Now it looks as if we are on the threshold of a new breakthrough—the atomic desalting of sea water in vast quantities for making the desert lands of this earth bloom for human need... Since we now know that the cost of desalting water drops sharply and progressively as the size of the installation increases, it is probable that sweet water produced by these huge plants would cost not more than 15 cents per 1000 gallons—and possibly considerably less... There is every reason to suppose that it could be a successful, self-sustaining business enterprise, whose revenue would derive from the sale of its products—water and electricity—to the users... The purpose... is... to promote peace in a deeply troubled area of the world through a new cooperative venture among nations.

These and other basic documents, including Strauss' memorandum outlining the proposal, have provided not only a flood of newspaper stories and magazine articles, but have also accelerated government-sponsored efforts. Of the engineering studies, two are directed specifically to foreign areas. Following President Johnson's meeting with Israel Prime Minister Levi Eshkol in June 1964, the Kaiser company was commissioned to make an engineering study of the feasibility of seawater desalting in Israel[2]. The Oak Ridge National Laboratory has produced a report on nuclear energy centers and agro-industrial complexes to be established in various arid areas of the world[3]. The tone of these two reports is cautiously optimistic, but a more careful review of their assumptions leads to quite opposite conclusions[4]. Others have written uninhibitedly—as if sweet water were already flowing into the desert at low costs[5].

The Kaiser study was specifically concerned with a desalting plant in Israel; the contemplated location of the plant was on the Mediterranean seashore about 9 kilometers south of Ashdod. The power plant would be a nuclear steam-generating facility using a conventional light-water nuclear reactor with a thermal rating of 1250 megawatts. Essentially all of the steam generated in the reactor passes through the generator without condensation; the steam exhausted from the turbine is condensed in the shell side of the brine heaters of the desalting plant. The evaporator structures consist of heat recovery stages, heat reject, and heat reject-deaerator stages. A multiple seawater intake structure would be located 450 meters offshore and 7 meters deep. An outfall facility would consist of a buried concrete box culvert with transition to an open channel beyond the desalting plant limits. The plant would have a capacity of 100 million gallons of desalted water daily; a plant operating factor of 85 percent is assumed. The generator would produce 200 megawatts salable electrical power at an estimated price of 5.3 mills per kilowatt-hour and an 85 percent power plant operating factor. Total capital costs are estimated from $187 to $210 million, depending upon interest rate. Annual operating costs vary from $16.8 to $28.7 million, for interest rates of 1.9 percent and 8.0 percent, respectively. Crediting power sales against total costs, water costs per 1000 gallons—conceived as the residual costs—range from 28.6 cents if interest is calculated at 1.9 percent to 67.0 cents if the interest is 8.0 percent. Several variations in structure and methods of operation are possible without major effect upon water cost.

The Oak Ridge proposal is for a major nuclear energy center, with industrial and agro-industrial complexes, as well as desalting works. It was conceived to be broadly suitable for several locations in the world, subject to specific site planning and adaptation. The proposal is based upon technologies expected to be developed over the next decade or two, not upon technologies tested and in application today—and that reason makes it less suitable for rigorous review since at *some* time in the future costs of both power and water production will presumably be lower than they are now. "Near-term" light-water reactors and "far-term" advanced breeder reactors were considered as well as near- and far-term desalting equipment; a number of alternative layouts were included, with industrial electrical power ranging mostly from 1585 to 2070 megawatts. Most of the designs would produce 1 billion gallons of desalted water daily. Investment costs in the nuclear plant and desalting works would range from $1.5 to $2.0 billion for most designs. Various industrial processes, producing metals or chemicals and using large amounts of electricity, are considered. Alternative costs are estimated, with different interest rates and other cost factors. A highly advanced type of agriculture is assumed. The Oak Ridge complex would produce ten times as much water as the Kaiser proposal, at a cost at "near-term" technology of 17 cents per 1000 gallons at 5 percent interest, 24 cents at 10 percent interest, and 32 cents at 15 percent interest and about one-third lower at "far-term" technology. As pointed out in the report, these values are arbitrary, since the complexes are conceived as closed economies. They represent the incremental cost of adding one unit of water to an existing plant. But, in the size class here contemplated, these incremental costs will approximate the average cost sufficiently to stand as surrogates.

Currently, a second stage of research, which includes the outlook for marketing the expected increase in output, has been undertaken at Oak Ridge to adapt the general design specifically to conditions as they exist in a number of locations in the Middle East.

Our purpose in this article is to explore the economic feasibility of desalting seawater on a large scale for commercial agriculture in regions of extremely low rainfall. In preparation of the material on desalting costs we have made extensive use of the analysis by Paul Wolfowitz[6] of the University of Chicago and a report by W. E. Hoehn of the RAND Corporation[7].

Little will be said here of the political aspects of such ventures except to note that we see no reason to believe that the desalting of seawater in the Middle East would have the peacemaking effects that have been claimed for it. The struggles between Israel and its neighbors have not been over freshwater, and even where water has been at issue, as in the case of the Jordan River, the differences could be resolved readily if there were peace in the Middle East, or at least an atmosphere in which negotiations were possible. Indeed, in 1955 these differences were resolved on a technical level, but formal agreement was frustrated by political antagonism[8].

Desalting seawater for large-scale commercial agriculture

Any program to desalt seawater for use in agriculture involves three closely interrelated components: (i) a source of energy; (ii) a process for producing sweet water out of seawater; and (iii) means for transporting the water, at the right time, to the place of its application for the growth of crops.

Each component is essential, and any part can set a technical or economic limit to the whole process. Much has been written about the first two components, but very little about the third.

Numerous sources of energy and methods of desalting exist, but we shall focus entirely upon nuclear power as the energy source and upon evaporation as the desalting method.

No persuasive case can be made for a preferred energy source. Indeed, one might simply stipulate a given cost of energy, from whatever source is locally most advantageous, and concentrate attention on the other two components. We are analyzing nuclear rather than fossil-fuel energy only because the proposals that have received most notice have been based on nuclear energy. First, the atom attracts both attention and funds. Second, there is a large well-funded atomic research establishment, certainly in the United States, staffed with imaginative and highly skilled thinkers who do not shrink from the novel and spectacular. Third, the larger the proposed installation, the greater the advantage of nuclear energy. And fourth, there are arid areas in the vicinity of seacoasts that are remote from other fuel sources and to which nuclear energy may offer less expensive access to the new technology of sweet water production. For these and perhaps other reasons the packages offered so far have contained nuclear energy as the energy source.

Regarding desalting, it may safely be assumed that current technology offers no more feasible way to obtain freshwater from the sea on a large scale. Even the so called "far off", 20-years-in-the-future, technology used in the more favorable of the two Oak Ridge variants is based on evaporation. That a breakthrough—say a very efficient, stiff, membrane—could change the picture goes without saying. But such breakthroughs are not now in view, despite much effort in that direction. Nor would they have a necessary advantageous association with power generation.

In short, we discuss the merits of the programs in the terms chosen by their proponents. Although costs would vary with the location of the plant and other environmental factors, it is possible to consider the problem in general and to reach conclusions which no specific application could significantly change. This is true even for the third component—the conveyance of the water once produced—provided that areas are eliminated which do not have suitable soils, or are too remote from the seacoast, or do not in other ways qualify.

Nuclear energy

The reputation of nuclear power as a cheap source of energy understandably has reached the desalting field, once it was realized that the addition of power production (from any fuel) to a desalting plant represented a logical combination, wherever raising of steam was part of the desalting process. Nobody now doubts that electricity from nuclear power plants can indeed be fully competitive with that from fossil-fuel plants under certain conditions in certain areas. But it is also true that some of the enthusiasm of recent years was based upon circumstances unlikely to be repeated in this country or to be found at all in less-developed regions of the world. These include the large funds furnished by the government for R&D, and the initial input made by suppliers who quoted highly attractive prices for their generating equipment when it seemed essential to the spread of the new technology. For these and other reasons, some sober criticism has been directed at the evaluation of the outlook for nuclear energy[7,9]. This does not cast doubt on its basic competitive position but does question the extent of this advantage. These are major considerations:

1) It will be a year or two before even one U.S. nuclear plant designed to be competitive with fossil-fuel plants has been in operation long enough to establish a performance record that would substantive expectations. The large scaling-up in size of the equipment ordered for nuclear plants in 1967 and 1968 and the reliance placed in all cost estimates on minimum downtime make this especially significant. What little experience has accumulated from demonstration plants in the United States indicates that the high rates of availability, a *sine qua non* of low power cost assumptions, will not be easy to attain.

2) The cost of nuclear generating equipment for the long run is far from settled. Past reductions in equipment prices by manufacturers have turned out to be more in the nature of initial lures. Costs in 1967 and early 1968 were $30 or more per kilowatt installed above those of 1965 and 1966, allowing for differences in size[7]. Nor has the trend abated[10]. Costs of conventional equipment have also risen, but less steeply.

3) There is some uncertainty regarding the future costs of nuclear fuel, once the increased power generation begins to reduce the uranium supply and forces a diversion to higher cost sources of the mineral. But prices of competitive fuels cannot be assumed as constant either, so that uncertainty is the real problem.

4) Nuclear power plants, owing partly to the heavy cost of shielding and containment, require more capital than conventional plants do[11]. Whenever a portion of the fuel is awaiting enrichment (or being enriched), being fabricated into fuel elements, awaiting loading, or undergoing cooling, it still represents capital investment and thus carries interest charges, no matter whether the utility or the supplier manages the fuel cycle. The recent steep rise in interest rates has been penalizing nuclear more than conventionally fired plants. No one knows whether, to what level and how soon, rates will begin to decline, but while rates are high they blunt the competitive edge of nuclear plants.

In addition to the factors cited which tend to make themselves felt in aggravated form in less-developed countries, there are some elements that apply especially to them.

Economics of scale

Electric power generation is a classic example of economies of scale: unit cost drops as size of the plant rises. This is especially marked for nuclear power plants, partly because the absolute cost of shielding and containment increases relatively little with reactor size.

Information available for 1967 shows a rise in capital costs per kilowatt capacity from about $130 for plants of 1200 megawatts to about $180 per kilowatt for plants in the 400-megawatt range. This explains why no nuclear power plant smaller than 450 megawatts has been ordered in the United States since 1963, and why the smallest size plant which a utility can now order from a major U.S. supplier is 480 megawatts[12].

However, these economies can be realized only when certain other favorable factors are also present. In the less-developed countries they are usually absent. Chief among them are a large market for electricity, that can accommodate a very large plant, and a well-developed power grid.

An engineering rule of thumb calls for reserve capacity equal to at least the largest single generator in the system to assure continued supply when that unit is out. In fact, to keep the system from collapsing in case the nuclear plant trips out, prudent engineers advise against installing in a small isolated system a nuclear plant that is larger than 10 percent of the peak load.

How many countries are there that fulfill the conditions which permit them to benefit from the economies offered by large nuclear reactors? To be competitive with power from conventional sources, "... a reactor of 500 Mw, now about the lower limit in size, would ... have to produce not less than 3.5 billion kwh per year. Presently, there is only a handful of countries outside of North America, Europe, and Oceania that consume that much electric energy per year *altogether*. And, of these, only Argentina, Brazil, Japan, India, North Korea, and the Republic of South Africa consume greatly more"[13]. Even though power markets will, of course be larger 10 and 20 years from now, it is precisely this circumstance, the "low-demand trap", that has led to the search for an adequate and reliable market, hence the recent work on agro-industrial complexes as built-in consumers.

Availability of capital

Because nuclear power plants require a large capital outlay per kilowatt of installed capacity, the availability of capital is of great importance. Most of the countries which could best use additional power—and water—are seriously short of capital; alternative investment opportunities exist which can earn interest at much higher rates than are customary in the United States even now. Israel, for instance, permits a legal maximum of 11 percent, and the demand for loans is usually greater than can be supplied at this rate[14]. Higher rates are paid in various ways. It is certainly doubtful if any country which could use a large desalting project based on nuclear power should count on having to pay less than 10 percent interest per year. If for political or other noneconomic reasons the United States should decide to provide a plant in a country on a subsidized basis, any interest rate could be used in the calculations. Without discussing the merits of subsidization, however, current efforts to portray nuclear desalting as having come or about to come "of age" are based not on subsidized but on market conditions.

Costs of equipment

Power plant costs would almost surely be higher outside the United States, especially in countries that have been mentioned as candidates for desalting plants[7] (p. 165). The reasons include costs of shipment of equipment, lack of supporting industries and their production, shortage of national specialists and construction crews, and longer construction time. Only a small portion of these increases might be made up by procurement of some of the equipment from lower cost sources abroad.

Operating performance

The cost of servicing a nuclear power plant is likely to be higher than in the United States for reasons very similar to those just cited. The intrusion of a highly complex technology into an environment that is not geared to it is bound to result in lessened effectiveness and higher cost of maintenance and operations generally. Although the record of power availability of nuclear plants is anything but good in the few plants that have so far operated in the United States, it is likely to be poorer in the less-developed areas of the world unless the plant can be run as a virtual enclave, and even then a good record is by no means certain. This is not to say that improvements would not gradually be attained, as they are bound to be attained in the United States. It is reasonable to believe that the economics of nuclear desalting, examined alone, would make the first plant ordered in any less-developed country disproportionately large, and it would be a long time before a second and a third could be built. Thus the initial plant would for years bear the burden and cost of serving both as an economic input and as a training and experimental facility.

We do not wish to appear unduly pessimistic. Not all of the adverse factors need come true, but some are sure to be felt. And there is little in the picture that points to the emergence of unforeseen favorable elements, at least not without consideration of other energy sources. A good deal of what has been said above would not be true of fossil fuels, abundantly available in the Middle East, where vast amounts of natural gas are flared, and where the marginal cost of crude oil is extremely low. Moreover, the economies of scale would be less pronounced and size problems somewhat alleviated.

But for the moment no such proposals have caught the public fancy, although there is no generally valid technical, economic, or other connection between nuclear energy and desalting. Indeed, from the viewpoint of international complications, the association of desalting with nuclear energy probably represents an obstacle rather than an aid to achievement of the economic objectives in some parts of the world. A first indication of change in that direction could be the proposed bill transmitted to the Senate by the Department of the Interior on 17 January 1969, 3 days prior to the Administration changeover. It would authorize U.S. participation in a dual-purpose plant to be erected in Israel. An upper limit of $40 million would be placed on any grant made to help finance the desalting techniques and necessary modifications in the power production of the plant. Financing of the balance as well as the choice of energy source would be determined by the Israeli government.

Efforts to overcome the difficulties briefly described above have taken two forms. One has been to present an optimistic picture of the expected costs of water by making highly favorable assumptions for cost and output factors. The other has been to broaden the scope of operations beyond production of power and water and to test the feasibility of a large agro-industrial complex in which the large volume of electricity that cannot be absorbed by the ordinary demand of the country can find a ready market.

First, Kaiser used plant costs based upon a price schedule that became quickly outmoded, as we have pointed out above, and has led to later revisions[15]. This would matter less if prices of crops had undergone similar increases, but they have not. Second, interest charges, and fixed charges based upon that interest, are unrealistically low. Even the highest variant has an interest charge of only 8 percent. The assumed downtime for the generating plant is 10 percent, certainly a highly optimistic assumption over the lifetime of the plant[6]. And since the cost of water is arrived at by deducting from the total costs of the dual plant the income from selling the large amounts of electricity that are not consumed internally, a constant price of that electricity is assumed for the lifetime of the plant. This method discounts the possibility of slowly falling revenue stemming from a declining power-cost level in the economy as a whole. To the extent that the income from electricity sales is maximized, the "cost" of water, as the residual, is minimized.

The Oak Ridge study does make some allowance for higher costs outside of the United States, and sets out a wide range of possible interest rates. But in terms of the Oak Ridge concept, there is no need (and perhaps no basis for doing so) to determine separately the cost of either power or water, since it is the returns for the operation as a whole that measure the profitability of this closed complex.

We shall deal later with the various assumptions, but one needs mentioning here. In both studies joint production economies are reflected in the cost of water. While such a subsidy from one part of joint production to the other may be wholly desirable in a given case, it is apt to mislead those who are interested in the cost of desalted water regardless of its association with power production. In fact, the popular discussion has fastened on precisely the costs of water that have emerged from such studies without awareness that water cost is to a substantial degree a function of the price at which power can be sold. In this connection, the Oak Ridge study straddles the fence. Although its basic concept of an integrated complex renders the costing of either power or water meaningless, it fails to exploit this advantage in a consistent way and presents both costs separately, albeit in a somewhat offhand manner, and, it must be said, to the decided detriment of the entire exercise; for it leads the reader to marvel at the agricultural sector calculations being based predominantly on water at 10 cents per thousand gallons, when the rest of the study clearly spells out that no such water is in the offing, not even in the "far-term" model 20 years hence.

The same phenomenon has turned up in the case of the dual plant in the Los Angeles area, discussed briefly below. Before it was tentatively shelved in mid-'68, the estimated cost of water at the plant had risen from 22 to nearer 40 cents per thousand gallons. But "20 cents per thousand gallons" has left a lasting impression with well-intentioned but ill-informed writers and speakers.

Desalting process

Much of the uncertainty to which we have drawn attention in the discussion of the energy-producing component is present in aggravated form in the desalting component. Here too, the scale of operations proposed in each instance is greatly in excess of anything that has so far been tried, although in the Kaiser proposal the large capacity is reached by replication of small, basic modules that are only five times the size of anything now in operation.

In each proposal the nuclear power plant and the distillation plant would be closely linked. Anything which led to a shutdown in one would force an early shutdown in the other, although planned maintenance in either process might be carried out during forced shutdowns of the other process. The schedule for a power and distillation plant in the Kaiser proposal calls for a demanding availability of 85 percent jointly, or a downtime of 15 percent. There is little on which to base an appraisal of this assumption. But it may be noted that the Point Loma demonstration plant of the Department of the Interior, prior to its transfer to Guantanamo Bay in Cuba, had an availability of only 70 percent[6]. Since this was an early plant, one would expect later ones to operate more continuously, if it were not for the type of difficulties that the Point Loma plant experienced. A serious one that the Kaiser plant does not seem to have adequately taken into account was the problem of drawing water out of the ocean. Many materials obstructed the intake pipe—kelp, sand and silt, fish, even large stones. Large and expensive stilling basins must be installed if such difficulties are to be avoided, and one may assume that the type of difficulty will vary from one location to another. "It is evident from our observations both at San Diego and elsewhere", an engineering evaluation states, "that the importance of trouble-free intake systems either does not get through to those responsible for the design of the system or that there exists a tendency to skimp in the design in order to reduce costs"[16].

Another unknown is the discharge of hot and bitter brines in volumes 100 times and more than that of existing plants. This could present awkward problems. At the minimum, the discharge point must be removed by considerable distance from the intake point to prevent even partial recirculation of ever saltier water; expensive piping out to sea may be required[17]. Adverse ecological consequences of dumping these wastes are inevitable. Neither they nor the possible costs of dealing with them have received attention.

Although each report is concerned with the future, some comparison with present plants is sobering. The lowest cost plant operating today (providing

17 Clawson/Landsberg (0271)

water for Key West, Florida) produces water for 83 cents per 1000 gallons, but with a subsidized interest rate loan from the federal government; without it the cost would have been very close to $1 per 1000 gallons. In 1966, two private utilities in Southern California, the City of Los Angeles, and the Metropolitan Water District of Southern California, assisted by funds from the U.S. Department of the Interior and the Atomic Energy Commission, entered into an agreement to build a desalting plant on man-made Bolsa Island in Southern California to serve an urban area with a high demand for both electricity and water. The plant was to produce 150 million gallons of desalted water daily and have a generating capacity of 1.8 million kilowatts of electricity. The originally estimated cost including water conveyance and power transmission was $444 million; by the summer of 1968 estimated costs had escalated to $765 million, owing to a greatly lengthened construction period, increased equipment and interest cost, stricter design criteria, and, one of the smallest items, a 10 percent larger power output. The project is uneconomic at this price, and the proposal in its present form has been shelved.

If desalting of seawater is not economic in Southern California today—where alternative water must be brought long distances at high cost, where electricity surely has a ready market, and where much of the water would not go to agriculture—then where is large-scale desalting of seawater economic? If it is not economic at an interest rate of 3.5 percent, and at the lifetime capacity factor for both water and power of 90 percent, assumed for this venture, then what are the prospects under less generous assumptions?

These recent experiences may not apply to desalting costs in the more distant future, but they are at least sobering. This is particularly true when one attempts to make corrections both for the unrealistically low fixed charges, and especially the interest rate (the Oak Ridge proposal is the most realistic in that respect), and some allowance also for the optimism incorporated into the estimates at various stages. Wolfowitz has tried to make adjustments for the proposed Israeli plant[6]. Using as his point of departure the lowest estimated cost of 28.6 cents, based on an interest charge of 1.9 percent, he demonstrates persuasively that the likely contingent expenses not included would bring the cost to 40 cents per 1000 gallons at the farm. If adjustment is then made to a more realistic but still modest fixed charge such ss 10 percent, the resulting cost of water at the farm would rise to somewhere between 90 cents and $1 per 1000 gallons.

Application of desalted seawater to the land

The third, and most generally neglected, aspect of desalting seawater for use in large-scale agriculture is the conveyance of the water from where it is produced at the edge of the sea to the land, which may be some distance inland and at a much higher elevation. The desalting plants discussed above will produce water in a constant stream (except when shut down for repairs or servicing), but the farmer wants water in a different time sequence during the year. In some way, water must be stored and transported, from one time and place pattern to another, and substantial costs will be incurred in doing so. Much of the discussion of the economics of desalting seawater overlooks this point; someone will compare the costs of water at the plant (usually grossly underestimated) with the value of the water at the farm (usually grossly overestimated).

In an arid region, irrigation water is essential for successful production of most crops, but so are several other inputs. The farmer combines them all into the farm operation program for production of crops and livestock which, in view of prices, costs, and markets, seems to him most likely to produce the greatest net income. The resulting time sequence of irrigation-water use is usually highly seasonal in character, its exact pattern depending upon climatic factors as well as upon choice of crops and methods of crop production. Modifying the farming program to smooth out the seasonal demand somewhat for irrigation water is possible in some areas and under some circumstances, but this modification is very likely to reduce income, sometimes substantially, from the whole farm operation. By and large, for desert and arid areas where desalted water might be used, a markedly seasonal demand for irrigation water is certain, if the farmer is free to choose when he takes water; demand for off-season water may be low.

The problem of storing and conveying water from desalting plant to farm will vary greatly from one location to another, but some generalizations may be made. Desalted water, in excess of immediate need, might be stored in surface reservoirs or underground aquifers located en route or not too distant from the place of either production or application, or in the soil of the farm. In each case, some water—often a great deal—will be lost through evaporation or percolation, or both; water stored in the soil may pick up salt—a great deal in most desert soils. Evaporation in most desert areas is high, often 10 feet or more annually from a water surface. There may be no suitable reservoir site; in any event, dams cost money to build. Soils and aquifers may have a low water-holding capacity or intake rate. Also, some means must be provided for carrying water by large conduits, pipes, or canals from the desalting plant or storage site to the border of each farm. In the United States, this has proven rather costly even when the water source was available by gravity flow. If the arable lands lie at some elevation, pumping costs will be considerable.

In the Kaiser report, the water-conveyance facilities and electrical transmission lines are not included. It is stated that they would add more than 15 percent to the investment. The water cost estimates are based upon 310 days annual operation of the desalting plant but no provision is made for storing this water at times of slack demand and no allowance is made for pumping costs. There are few good surface reservoir sites in Israel. The same limestone formations which allow infiltration of natural precipitation that could later be salvaged as groundwater also are the cause of leaky reservoirs. The most suitable lands in Israel near the proposed desalting plant lie at an elevation of 500 feet or more; pumping costs, even with relatively cheap electricity, would be considerable. The cost of taking desalted water from the plant to the field includes (i) losses in transport; (ii) pumping costs; and (iii) costs of conveyance to the farm, including distribution canals or pipes. By far the greatest of these is likely to be water loss. A 10 percent loss of water would raise the cost of the remaining water by 11 percent, a 20 percent loss by 25 percent, and a 30 percent loss by 43 percent. The more costly the desalting process, the more costly the loss of water in storage or in conveyance.

Pumping costs depend primarily upon lift and distance. Even with high pump efficiency, lifting water requires somewhat more than one kilowatt-hour for each foot of lift for an acre-foot of water (enough water to cover an acre one foot deep, or 326,000 gallons). A 500-foot lift, as would be necessary at the most frequently mentioned Israeli site, would require about 640 kilowatt-hours of electricity; at 5.3 mills per kilowatt-hour, the rate at which the Kaiser report estimates electricity can be disposed of, this would still mean nearly $3.50 per acre-foot for energy; depreciation, maintenance, and interest on pumping equipment would probably add as much again. Finally, there are the costs of construction, maintenance, and operation of a canal or pipe system. The annual cost, including interest on capital, could hardly be less than $3 per acre-foot.

The Kaiser report, on the basis of 8 percent interest on invested capital, arrives at a cost of 67 cents per 1000 gallons at the plant, or $218 per acre-foot. On the basis of the foregoing calculations, an overall loss of water of 10 percent (representing a much higher loss on the volumes actually stored), plus the other costs, would add about $34 per acre-foot to the cost, or 14 percent. If the overall loss were 20 percent, the lost water would add $55 per acre-foot to the cost of the delivered water; with the other costs, total costs incurred between distillation plant and field would be $65, or a 30 percent increase.

If all calculations in the Kaiser report were retained, but the interest rate raised to 12 percent, the costs of desalted water would be in excess of 75 cents per 1000 gallons. If 20 percent were added for conveyance costs and losses, the delivered cost at the farmer's field, on the time schedule he wants the water, rises to 90 cents or more per 1000 gallons.

If one accepts the Oak Ridge calculations, but uses an interest rate of 12 percent, the cost of desalted water at the plant is 28 cents per 1000 gallons; if 20 percent were added for conveyance costs and losses, the delivered price becomes 34 cents; and taking into account all the variables discussed, it seems realistic to count on a delivered cost of at least 40 cents per 1000 gallons, or $130 per acre-foot. It should be noted that some of these additional costs, here incorporated in the cost of irrigation water, are allowed for in various ways in the Oak Ridge scheme under various capital charges of the farm enterprise. Thus, comparisons are difficult because the cost of the water remains unchanged from its cost at the outlet of the desalting plant. But primarily, it is larger size and assumptions of less costly future technology that explain the lower costs of the Oak Ridge study as compared to the Kaiser study.

Value of irrigation water

The value of water for irrigation, whatever its source, is affected by many variables—climate, soils, associated inputs such as fertilizer, markets, efficiency of farmers, competition from other producing areas, and many others. Throughout the whole world, water is rarely sold on a market, hence one must estimate "shadow prices" for the irrigation-water supply. It is extremely difficult to determine the *actual* value of irrigation water, but not difficult to say how it should *not* be determined.

First, the value of irrigation water to be developed by the two desalting projects cannot be determined on the basis of what a few farmers could pay to produce a highly specialized crop for a special market. There has been much loose talk about production of "winter vegetables", for instance; aside from the fact that this type of agriculture has never been the gold mine that some think it is, and that competition among producing areas in the future will reduce whatever large profits may have existed in the past (it is hardly legitimate to assume that the advantages of new technology will not be available to other, similarly situated areas), the scale of the Kaiser and Oak Ridge projects preclude this type of agriculture for more than a small fraction of the water to be produced. One hundred million gallons a day for 310 days in the year—85 percent availability—in the Kaiser project, are nearly 100,000 acre-feet annually, or irrigation water for perhaps 35,000 acres of summer crops and much more of winter crops; the Oak Ridge project is ten times as large. Even 35,000 acres is not much less than the total acreage of all vegetables grown annually in Israel, of which only a small fraction are exported. Such an acreage of winter vegetables could not be grown at any single location for the home market and if exported would have disastrous results in terms of prices of products. True, tomatoes—greatly desired as a leading export—are grown in Egypt on some 200,000 acres but exports in 1965 were the equivalent, at prevailing yields, of the harvest from 40 acres! Even in 1960, the best recent export year, exports came from the equivalent of 700 acres. The task of escalating from such levels to those appropriate to the magnitude of the desalting plants is truly overwhelming. Such comparisons and our ignorance concerning the characteristics of the specialty markets lead one to conclude that crop production from large-scale desalting works must be primarily staple, not specialty, crops.

Second, one cannot safely assume that all the increase in value of output resulting from irrigation will, or can be made to, accrue to the irrigation water; this is a trap into which economists around the world have fallen repeatedly. The quality of the labor and the management which will be required under the more intensive irrigation farming will demand, and can get, higher returns than the kind of labor and management which sufficed for the less intensive agriculture that the new irrigation replaced. Moreover, farmers and other landowners the world over have demanded and have secured some part of the increased product resulting from irrigation as a reward for their land. Further, to attract the capital needed for the new irrigated agriculture, adequate rewards must be in prospect, including a generous allowance for risk. Some of the farm programs or budgets prepared for proposed new irrigation seem to show that very large sums can be paid for irrigation water. On closer examination, these have a fatal flaw; if the intended crop production is so profitable that very large sums can be paid for water, then it is profitable enough so that other extensive areas of the world, including those that need not pay high prices for water, can undertake such production—and the estimated price then quickly drops. Furthermore, the costs of other inputs rise rapidly, as the high yields conventionally assumed on irrigated acreage in these studies demand greatly increased applications of fertilizer, pesticides, and so forth, with attendant employment of sophisticated skills and machinery. In irrigated cotton-growing in California, for example, other costs are so high that water costs typically constitute only 10 to 15 percent of total operating cost.

Third, it is easy to develop plans which embody a wholly new order of magnitude in farm efficiency—crop yields much higher than those obtained by farmers in other irrigation projects in the region, fertilizer inputs several times as great as now practiced, new crop varieties that lead to much higher yields, and many others. By comparing irrigation agriculture on this new higher plane of efficiency with nonirrigated agriculture (or even with present irrigation) on the older, lower level of efficiency, some very high values of water can be estimated. Irrigation does indeed open up new production opportunities, but realism is called for in estimating just how much advantage can and will be taken of those opportunities, and how soon. If the new system of agriculture is possible with new irrigation, why is it not feasible with old irrigation? What reason is there to expect that provision of irrigation water will immediately transform a backward, traditional agriculture into a modern or futuristic efficient one?

Fourth, the agro-industrial complex has been offered as an answer to the last question asked. But is it? Such complexes as sketched by their proponents employ currently unknown or untested methods in industry and agriculture, produce for unspecified markets, and appear to justify very high costs for irrigation water. The prime example here is the Oak Ridge project. Although comprehensive in the scope of things to be considered, it tends to assume optimistic outcomes, uses low costs, and fails to allow for unexpected difficulties and costs. Above all, it fails to supply a satisfactory answer to this question: if these great agro-industrial complexes are economically feasible with desalted water, why are they not feasible with natural flow or groundwater? There is nothing magical about desalted water; it is simply water.

The agro-industrial complexes of the Oak Ridge type have been defended on the ground that they would constitute a new order of technology and organization, freed of all the inhibitions of restrictive institutions, cultural values, modes of living, and so forth, which impede agricultural and industrial development in some countries. This is a dubious argument if applied to Australia, Israel, and possibly to Mexico and India. Moreover, this proposal is futuristic plantation philosophy. In many colonies of the world before

World War II, there were plantation economies, using outside capital, outside management, and producing for an export market; often they were highly efficient. Most are now liquidated as foreign enterprises; there is little reason to expect that the countries would welcome them back. The very isolation of the proposed agro-industrial complex from the mainstream of the country's culture is its most devastating weakness, regardless of the efficiency it might attain. The Oak Ridge study comments on this by contemplating that the food factory concept "... would appear to be the reverse of agrarian reform programs in many countries. On the other hand, setting up an operation in a sparsely populated area might be effective in avoiding complications of existing social organizations and customs"[3] (p. 27). One can only comment that it would save even more trouble if one were to select a less difficult geographic, social, and political setting and then find a way of letting the country to be aided share in the fruits of production by assigning to it the plant's net return.

But ignoring these broader-based considerations and insisting only that the large-scale desalting projects planned by Kaiser and Oak Ridge must produce predominantly staple crops, such as grains and cotton, for domestic and export markets, one can judge the economic feasibility from a number of recent American studies that provide estimates of the value of irrigation water for such crops. Since the contemplated farming ventures discussed above are based on highly advanced technologies and must to a large extent be competitive with world market prices, such studies are not as inappropriate a criterion as one might first think.

Young and Martin provide information and analyses to indicate that the value of irrigation water in central Arizona is less than $21 per acre-foot[18]; Stults, considering the situation in Pinal County, Arizona, makes analyses which imply that the value of the water is about $9 per acre-foot[19]; Grubb estimated the ability to pay for irrigation water in the High Plains of Texas ranged from $27 to $36 an acre-foot, even in 1990[20]; and Brown and McGuire found that the marginal value productivity of irrigation water in Kern County, California, was about $19 per acre-foot[21]. These are all in fairly good farming areas, where the growing season is rather long, cropping patterns can be rather intensive, and crop yields are relatively high. In irrigated areas where farming is somewhat less intensive, due in part to differences in climate, Hartman and Anderson concluded that the value of supplementary water was from $1.50 to $3 per acre-foot[22]; and Fullerton found that in a fairly active water-rental area, the price was about $8.75 per acre-foot[23]. All of these examples involve rather high-level managerial competence (which is more easily hypothesized) unlike that found in some of the countries under study; the same is true of the availability of farm machinery, fertilizer, insecticides, and other inputs. It is important to note that they do not focus on the subsidized price of water, but on what users can afford to pay. Thus they are directly relevant to the hypothetical cost of desalted (or any other) water. Moreover, they escape the frequent criticism that the cost of desalted water should not be compared with the actual price currently paid for water, or that the present price of water is an irrelevant object of comparison, since it must be judged in a multi-purpose use context.

On the basis of this range of American experience, it seems most unlikely that irrigation water delivered to the farm on the schedule the farmer wants it, for the production of staple crops, can attain a value greater than $30 per acre foot (10 cents per 1000 gallons), and a value of $10 per acre foot (3 cents per 1000 gallons) is a much more reasonable planning standard.

The conclusion is inescapable: the full and true costs of the proposed desalting projects, now and for the next 20 years, are at least one whole order of magnitude greater than the value of the water to agriculture. The specifics of both cost and value will vary, depending upon the location of the plant and the myriad of factors associated with that location, upon what desalting costs actually are in practice, upon crop possibilities (costs and markets, especially), and upon other variables. But it is impossible to bring planned costs and prospective values for agriculture together or even close.

Nothing we have said with regard to the prospects for desalting seawater should be construed as an argument against continued research, including the construction of a rather large pilot plant. The Oak Ridge study both merits and needs attentive reading and critical review. Such research must not stop at the farm gate nor bypass the broader implications of such programs with a few passing sentences. There is more involved here than either "truth in advertising", the discovery of a new input, or a new means of fighting hunger. The present mirage may indeed have an oasis within it, and we as a nation have the resources to pursue the matter much further. But let us not delude ourselves or the rest of the world that an early and practical solution is at hand.

References and notes

1a. *Science*, June 6, 1969. Reprint is article in its entirety. Copyright by the American Association for the Advancement of Science.
1. D. D. Eisenhower, *Reader's Digest*, June 1968. By "huge" plants, he means a billion gallon daily capacity.
2. Kaiser Engineers and Catalytic Construction Co., *Engineering Feasibility and Economic Study for Dual-Purpose Electric Power-Water Desalting Plant for Israel*, January 1966. The study has been revised twice since, and may be secured from Kaiser Engineers, Oakland, California.
3. *Nuclear Energy Center, Industrial and Agro-Industrial Complexes*, ORNL-4290, UC-80-Reactor Technology (Oak Ridge National Laboratory, Oak Ridge, Tenn., November, 1968). The full report was not available at the time of writing, but has since been published.
4. See our letter to the editor, *Environ. Sci. Technol.* **2**, 648 (1968).
5. An example of this sort of writing is an article by V. Nikitopoulos, *Ekistics* **26**, 14 July 1968), in which the author presents a map showing the land areas of the world which *cannot* be served by desalinated water, namely those more than 1000 kilometers from any ocean.
6. P. Wolfowitz, *Middle East Nuclear Desalting: Economic and Political Considerations*, RM-6019-FF (RAND Corp., Santa Monica, Calif., 1969).
7. W. E. Hoehn, *The Economics of Nuclear Reactors for Power and Desalting*, RM-5227-PR/ISA (RAND Corp., Santa Monica, Calif., 1967).
8. G. G. Stevens, *Jordan River Partition* (Hoover Institution on War, Revolution, and Peace, Stanford Univ., Stanford, Calif., 1965).
9. See P. Sporn in *Nuclear Power Economics—1962 through 1967*, report of Joint Committee on Atomic Energy, U.S. Congress (Government Printing Office, Washington, D.C., 1968), p. 2.
10. The Atomic Industry Forum ["The Nuclear Energy Industry—The U.S. Highlights of 1968" (1968), mimeographed] puts the case even more strongly: "The direct costs of constructing nuclear generating plants rose significantly in 1968. From a low in 1966 of about $100 they had increased some 30–40 percent in 1967, and there seemed to be a strong consensus that this year's increase was also 30–40 percent. While the costs of comparable fossil-fueled units also rose, the increase was apparently less abrupt."
11. Interestingly, the same is true for the desalting phases. The "far-term" technology (combined flash-vertical-tube) requires more capital than the "near-term" (multistage flash) does.
12. The pessimistic outlook for the emergence of smaller but still low-cost reactors was presented in a paper given at the 1968 World Power Conference in Moscow, and entitled "Prospects for Small- and Medium-Sized Nuclear Reactor Plants", by W. Buenlich and P. H. Kruck (Central Office of World Power Conference, London, England).
13. H. H. Landsberg, in *World Population—The View Ahead* (International Development Research Center Ser. No. 1) R. N. Farmer, J. D. Long, G. J. Stolnitz, Eds. (Bureau of Business Research, Indiana School of Business, Bloomington, 1968), p. 138.
14. *The Economist* (London), "Quarterly Economic Review: Israel", annual supplement, 1966 [cited in Wolfowitz (6)].

15. For a detailed review of the Kaiser Engineering proposal see Hoehn (7). We have not so far seen any similarly careful review of the Oak Ridge study.
16. A.C. Foster and J.P. Herlihy, "Operating Experience at San Diego Flash Distillation Plant", in *Proc. First Int. Symp. Water Desalination*. (Government Printing Office, Washington, D.C., 1965 [cited by Wolfowitz (6)].
17. S.T. Powell, "Factors Involved in the Economic Production of Usable Fresh Water from Saline Sources", in *Proc. First Int. Symp. Water Desalination* [cited by Wolfowitz (6)].
18. R.A. Young and W.E. Martin, *Ariz. Rev.* **16**, 9 (March 1967).
19. H.M. Stults, *Water Resources and Economic Development of the West, Rep. No. 15, Conf Proc.* of the Committee on the Economics of Water Resource Development of the Western Agricultural Economic Research Council, 7–9 December 1966, Las Vegas, Nevada.
20. H.W. Grubb, *Importance of Irrigation Water to the Economy of the Texas High Plains, Texas Water Development Board Rep. 11*, (Austin, Texas, January 1966).
21. G.M. Brown and C.B. McGuire, *Water Resources Res.* **3**, 33 (1967).
22. L.M. Hartman and R.L. Anderson, *J. Farm Econ.* **44**, 207 (1962).
23. H.H. Fullerton, "Transfer Restrictions and Misallocation of Irrigation Water", thesis, Utah State University (1965).

INDEX

Acid scale control   18
Agriculture   4, 6, 36, 165, 200–1, 202, 204, 251 ff
  carbon dioxide level   41
  productivity   41, 73–4, 240–6
  see also Crops, Irrigation, Water
Agro-industrial complex   4, 7, 94–5, 97, 112, 114–21, 154, 162 ff, 218, 247, 252, 253, 257, 259, 261
  location   119, 247
Algiers   81
Amortization   42, 54, 139, 152, 155
Annual fixed charges *see* Charges
Application, commercial   11
  plant, by process   14
Applied Research and Engineering Ltd (AREL)   147 ff, 222 ff
Aqaba, gulf of   81, 89
Aqueduct   27, 35, 36, 239 *see also* Water conveyance
Arabian peninsula   3
Ashdod (Israel)   46, 50, 138 ff, 252
  plant case study   138–46, 252–3
  plant water cost   139
Atomic energy   68 *see also* Nuclear
Atomic Energy Commission (U.S.) (AEC)   4, 44, 72, 77, 134 ff, 175 ff, 190, 210, 251

Baja California (Mexico)   21, 35, 83, 192 ff, 239
  agreement (U.S.-Mexico-IAEA)   192
  basic rules and program   193–5
  dual-purpose scheme   192–206
  study analysis
    anlasserved, power and water   200 to 201
    economics   198, 200, 202, 204, 205
    costs   203, (updating) 205
    phasing   197–8, 203, 205
    power needs   201, 202, 204
    power transmission   201

Baja California (Mexico)
  study analysis—*cont.*
    technical concepts   202, 205
    water demands   195–6
    water marketing   199, 200, 203–4
  study team   192–3, 217
Baker, Senator Howard H.   62–3, 64–5, 66, 207
Bechtel Corporation   177 ff, 181, 222 ff
Bolsa Island (U.S.A.)   4, 5–7, 76–7, 83, 84, 171–91, 262
  project organization   172–3, 177–80, 188
Brackish water   11, 19, 20, 22, 23, 69
  Test Center   23
Brine discharge   261
Browics Ferry reactor station   247
Bureau of Mines   73
  of Reclamation   69, 74, 75, 137

California   3, 5, 7, 21, 125
Capacity factor   131, 143–4
  see also Plant
Capital investment   35, 36, 257 ff
Carbon dioxide   41
Catalytic Construction Co   44, 58, 222 ff
Central Rice Institute (India)   242
Charges *see also* Costs
  annual fixed   15, 16, 47, 48, 53, 54, 57, 124, 135–6, 140–6, 203, 227, 253
  capital   15, 17, 18, 47, 48, 116, 256
Chemical dosing   19
Clair Engle (San Diego) plant   22
Cloud seeding *see* Weather modification
Coastal desert communities   35–40
Colorado aqueduct   238
  River   5, 35, 174, 192, 196, 200
COMSAT   65
Concrete   23, 56
Construction, plant, by process   13
Corrosion   19, 23, 34

272

Index

Costing case studies 132–67
  methods 125–31
Costs analysis 16–17, 42, 47, 48, 116, 141–170, 203–4, 221 ff, 252–4, 258, 262 ff
  annual 15, 17, 18, 58, 60, 118, 203
  capital 15, 16, 45, 46, 48, 54, 58, 116, 118, 120–1, 140 ff, 203, 221 ff, 256
  comparative 1, 59, 60, 128–131, 138–170, 221 ff, 243 ff, 255–6
  escalation 182–8, 216–7, 219, 221 ff, 244, 262
  estimated 41, (case studies) 132–170
  incremental operating 128–131, 253
  operating 15, 17, 47, 49, 54, 56, 116, 121, 253
  parametric estimation 146–154
  prospective 122–170, 203–4, 210
  unit water 15, 16, 17, 26–7, 36, 45, 46–7, 48, 52–5, 58, 60, 88, 104–5, 123 ff, 203, 209, 227 ff, 253, 260, 261–5
  see also Ashdod, Baja Calif, Bolsa Island, Plant, Power, Pumping
Credit methods (costing) 126 ff
Crops, water-yield relationship 112–4, 214–3
  water requirement 240 ff
  see Agriculture, Irrigation, Water

Demonstration plant 210
Department of the Interior 3, 4, 44
  see Udall, A, E, C, OSW, Reports
Water and Power (Los Angeles) (DWP) 175 ff
Desalination
  agriculture 240 ff see Agriculture, Agro-Industrial, Irrigation, Water
  large scale 6, 21, 26, 30, 208–210, 214, 219
  nuclear 210, 217, 219 see Dual-purpose, Nuclear
  output 2, 22, 46, 259
  plant 1, 10, 17, 253 see Plant
  prospects 2–8, 20, 76, 246, 254 ff
  systems see Processes
  technology 20
Desalted seawater see Seawater

Desalting see Desalination
Design 50
  conceptual 49, 51, 57
  criteria 185
  ideal 16
Distillation see Processes
  condenser design 18
Distribution 54
  plant, geographical 12
Dual-purpose water-power plant 3, 4, 5, 11, 17, 21, 44, 46, 50–7, 70, 83, 84, 123, 126 ff, 140 ff, 168, 178, 188, 192, 201, 209, 213, 215, 230, 246, 260
  fossil fuel 45, 47, 53, 70, 84, 129–130, 140, 168, 230, 255
  integrated production and use 114–121
  nuclear 45, 47, 49, 53, 84, 129–130, 140, 171, 252
  see also Baja Calif, Bolsa Island

Economic analysis for profitability 103
  feasibility 71–2, 210
  see Costs
Economies of scale 257, 259
Eisenhower, General D.D. 63, 68, 88, 89, 156, 158n
  statement 66–7
Electric power 4, 38, 46, 128
  grid 73–4, 129
  power capacity rating 128–9
  see Dual-purpose, Nuclear, Power
Energy center concept 218–9
  centered complexes 96, 214
  see also Agro-industrial, Nuclear
Estimates see Costs

Fabrication, field see Plant construction
Feather River 35
Feedwater effects of 32–4
Florida City 27
  Keys Aqueduct Commission 27
Fluor Corp. 27, 76
Foreign Relations Cttee see Senate
Fossil fuel 97–8, 124, 128
  see Dual-purpose, Key Ivest
  boilers 123

Freeport Test Bed Facility (Texas) 23
Freezing process 14

General Electric Corp. 140ff
Geological survey 73
Grain production cost 243–6
Greenhouses inflated 40
Gore, Senator 62, 70, 72, 78–79, 81–2, 85

Heat exchanger 40
  high quality 11
  low quality 11
  recovery equipment 38
  see Water temperature, Nuclear
Hydrochloric acid 33
Hydro-electric power 120, 129, 213
Hydrogen sulfide 30
Hyperfiltration process see Reverse Osmosis

India, Ganges Plain 96
  Tarapur 140
  Trombay 96
Industrial complexes 218, 253
  Cost comparisons 108, 110–11
International Atomic Energy Agency (IAEA) 21, 81, 89–91, 92, 123, 125, 127, 147ff, 192, 207, 210, 217
Iran 69, 73
Irrigation 22, 83, 240–6, 265ff
  overhead 81–2
  returnflows, depollution 23
  see also Water
Israel 3, 44–58, 75, 125, 132, 138ff, 212, 259, 264
  Atomic Energy Commission 44
  Electric Corp. 44, 46, 48, 49, 54
  see also Ashdod, Aqaba

Johnson, Lyndon B. (President) 3, 25, 44, 68, 70, 180
Jordan 69, 75, 254
  River project 75, 87–8

Kaiser Engineers 7, 44, 58, 78, 138ff, 222ff, 252ff, 259, 264
Key West, Florida 3, 22, 26–34
  plant cost data 33, 226ff
Kuwait 11, 69, 73

Lebanon 69, 75, 88
Limited energy center 218
  see Energy
Load factor, peak-month 16
  plant 15, 16, 17
Los Angeles Dept of Water and Power 21
  dual-purpose plant 260

Materials Test Center (Texas) 23
Metropolitan Water District (S. Calif) (MWD) 21, 77, 132ff, 172ff, 223ff, 238
Mexico 3, 5, 21, 192–206
  Govt. 192ff (Agreement)
  study 226
Middle East Hearings 62–92
  proposals 154, 161, 166
Morocco 81
Multiple reactor stations 104

National Park Service 73
Nitzanim (Israel) 46
North American Water and Power Alliance (NAWAPA) 43
Nuclear power 4, see also Desalination, Dual-purpose, Energy center, Reactor
  desalting, case for 208–220
  energy centers 35, 93–121, 208, 210, 214, 252, 253
  energy centers study areas 100–101
  energy cost 208–9
  plants 44–57, 84, 207, 214, 256, 258
  reactors 123, 246
  steam systems comparison 128–31

Oak Ridge National Laboratory (ORNL) 4, 7, 35, 43, 72, 86, 96, 121, 154ff, 166–7, 210, 214, 218, 222ff, 248, 252, 253ff
Office of Saline Water (OSW) 20, 25, 36, 44, 70, 73, 76, 134, 176ff, 210, 226, 251
Osmosis, reverse see Reverse

Persian Gulf 3, see Kuwait
Phosphate dosing 18

Index

Planning cycle 219
Plant capacity 11, 13, 16, 22, 26, 46, 50, 52, 53, 56, 58–61, 70–71, 84, 253
  comparative costs 221–234
  cost 26, 33, 50–52, 56, 88
  design 28
  field construction 29, 52
  growth, controlled environment 37, 38–43
  life 131, 136, 141
  location 46, 50, 51, 107, 186, 198–9, 203, 213, 269
  operating factor 45
  operational problems 19
  performance 9, 258 ff
  shop-fabrication 52
  viability 213–215
Point Loma plant 261
Pollution 238
Power comparison of processes 106
  consumption 50, 209
  costs 104, 107, 109, 123 ff
  credit 48, 51, 53, 54, 59, 128–131, 140 ff
  demand elasticity 93
  marketing 201, 205, 253, 257, 260
  production 38
  use of 105–112
  see also Electric, Nuclear
Process, geographical distribution of plants by 12
Processes, distillation/evaporation 23
  electrodialysis 11, 16, 17
  freezing 14
  humidification 38
  long-tube vertical (LTV) 11, 116, 202, 209
  multi-stage flash (MSF) 16, 17, 18, 22, 26, 28, 45, 49, 50, 52, 116, 173, 209
  reverse osmosis 14, 22, 23, 255
  single-unit flash 26, see Key West
  submerged-tube (HMSE) 11, 16, 17
  vapour compression (VC) 11, 16, 17
  see Dual-purpose, Nuclear, Reactor
Productivity rates 43
Project economics 213–4
  financing 214–5
  MEND 64–5

Project economics—*cont.*
  organization 215–6, 217
  phasing 216–7
  preparation 211–13
Proration methods (costing) 126
Puerto Peñasco (Mexico) 3, 35
Pump, brine recirculation 22
Pumping cost 264

Questionnaire, desalination *see* United Nations

Ramey, James T. (Commissioner AEC) 5, 7, 72, 77, 85, 96, 169–170, 207–220
Rand Corp. 4, 121, 254
Reactor boiling water (BWR) 49, 50, 51, 53, 59
  comparative characterstics 60, 128 ff
  pressurized water (PWR) 49, 50, 51, 53, 56
References (sources and reports) 19, 25, 34, 43, 61, 92, 121, 170, 191, 206, 220, 235–7, 249–250, 270–271
Reports 2, 101
  United Nations (Dept of Economic and Social Affairs) 2, 7, 9, 10
  U.S. Govt. (Dept of Interior)
  1968 Saline Water Conversion 20 ff
  see also Baker 64–5
  Eisenhower 66–7
  Kaiser 7
  Udall 3, 67–84
  and footnotes 9, 10, 96–99, 123, 125, 127, 132, 134–140, 142–8, 151, 154–7, 159, 162–5, 167, 223, 227, 234
Resource management 2
Reverse osmosis *see* Processes
Rice 241 ff

Saline Water Conversion, Act (1958) 24
  Program (1970) 24–25
Salt production 11
San Diego Gas and Electric Co 21, 175 ff
  Test Facility 22, 234
San Luis Rio Colorado 21
Santa Clara 21
Saudi Arabia 69, 70, 81

Index

Scale 135
  economies of 257–259
Scale Control 18, 19, 38, 209
Scaling properties 55
Seawater desalted 1, 4, 21
  intake 261
  use of 35, 254–5
  see Desalination
Senate (Foreign Relations Commitee)
  Resolution 155 (1967), 4, 62–3, 167, 251–2
  Hearings 7, 62–92
Seismic forces 186, 205, 217
Sewage 49
Shevchenko (Caspian Sea) 239
Sinking Fund *see* Amortization
Siting *see* Plant location
Solar energy 36
Sonora (Mexico) 21, 83
Sources *see* References, Reports
Southern California Edison 21, 77, 175 ff
Soviet Union 72, 165, 239
Stanford Research Institute 23
Stock Island 27, 29
Storage 19, 263 *see* Water
Strauss, Adm. L.L. 63, 67, 85, 156
Strauss–Eisenhower Proposal 154, 156 ff, 163 ff, 252
Syria 75

Tahal (Israel) 44, 46, 49
Tarapur (India) 140–141
Time discounting (cost) 130
Treatment of raw water 23
  sewague 23
Trombay 96
Tubes, fluted and spiral 23
Tunisia 81
Turbines 28, 29

Udall, Stewart L. (Secretary of the Interior 3 (report) 20–25, 67–84 (statement) 85, 190

United Nations (Dept of Economic and Social Affairs)
  Desalination Questionnaire 10
  First Desalination Plant Operation Survey 9 ff
  Seminar–Economic Application of Water Desalination (1965) 9
  see References, Reports
United States Govt. 21, 172, 192
  see References, Reports
University of
  Arizona 36, 38
  Chicago 254
  Sonora 36, 38
Uranium prices 130
Use of water 112–114
  basic comparison 1, 35
  see Water

Variants, effects of 55, 56

Waste thermal energy 36
  land 120
Water agricultural 251 ff, 263–9
  brackish 11, 19, 22, 23, 69 *see* Brackish
  consumption 49, 238
  conveyance 46, 198, 204, 223, 254 ff, 263–5
  ground 76, 165
  irrigation, value of 265–9
  needs 199, 200, 202, 203
  production 47, 50, 239
  quality 19, 49–50, 205, 209, 214
  re-use 76
  storage 19, 263
  temperature 18, 19, 22, 28, 29, 38, 40, 45, 49, 52, 55, 60, 123, 148, 174
Water Resources Council (U.S.) 24
Water-power *see* Dual-purpose
Weather modification 75
Weinberg Dr 86, 92, 93, 164
Westinghouse Electric Corp. 3, 26–7, 226

Young, Gale (Asst Director ORNL) 6, 7, 238–250